Culture

Culture

Leading Scientists Explore Societies, Art, Power, and Technology

Edited by John Brockman

HARPER  PERENNIAL

NEW YORK • LONDON • TORONTO • SYDNEY • NEW DELHI • AUCKLAND

HARPER ● PERENNIAL

HarperCollins books may be purchased for educational, business, or sales promotional use. For information please write: Special Markets Department, HarperCollins Publishers, 10 East 53rd Street, New York, NY 10022.

FIRST EDITION

Library of Congress Cataloging-in-Publication Data

Culture : leading scientists explore civilizations, art, networks, reputation, and the on-line revolution / edited by John Brockman.
p. cm.
Summary: "A short, cutting-edge master class covering everything you need to know about culture. Edited by John Brockman, with contributions by the world's leading thinkers"—Provided by publisher.
Includes bibliographical references and index.
ISBN 978-0-06-202313-1 (pbk.)
1. Civilization, Modern—21st century. 2. Culture. 3. Popular culture. 4. Art and society. 5. Social networks. 6. Reputation. 7. Technological innovations—Social aspects. 8. Internet—Social aspects. I. Brockman, John, 1941–.

CB430.C88 2011
909.83—dc22   2010052195

11 12 13 14 15 OV/RRD 10 9 8 7 6 5 4 3 2 1

Contents

Introduction

John Brockman

Editor and publisher, Edge.org

In summer 2009, during a talk at the Festival of Ideas in Bristol, England, physicist Freeman Dyson articulated a vision for the future. Responding to the recent book *The Age of Wonder*, in which Richard Holmes describes how the first Romantic Age was centered on chemistry and poetry, Dyson pointed out that today a new "Age of Wonder" has arrived that is dominated by computational biology. Its leaders include genomics researcher Craig Venter, medical engineer Dean Kamen, computer scientists Larry Page and Sergey Brin, and software architect and mathematician Charles Simonyi. The nexus for this intellectual activity, he observed, is online at *Edge* (www.edge.org).

Dyson envisions an age of biology in which "a new generation of artists, writing genomes as fluently as Blake and Byron wrote verses, might create an abundance of new flowers and fruit and trees and birds to enrich the ecology of our planet. Most of these artists would be amateurs, but they would be in close touch with science, like the poets of the earlier Age of Wonder. The new Age of Wonder might bring together wealthy entrepreneurs like Venter and Kamen . . . and a worldwide community of gardeners and farmers and breeders, working together to make the planet beautiful as well as fertile, hospitable to hummingbirds as well as to humans."

Indeed, Dyson was present at the August 2007 *Edge* meeting "Life: What a Concept," where he and genomics researchers Craig Venter and George Church, biologist Robert Shapiro, exobiologist and astronomer Dimitar Sasselov, and quantum physicist Seth Lloyd presented their new and, in more than a few cases, startling research and/or ideas in the biological sciences. "The meeting," according to *Sueddeutsche Zeitung*, the largest national German newspaper, "was one of those memorable events that people in years to come will see as a crucial moment in history. After all, it's where the dawning of the age of biology was officially announced."

So, what is *Edge*?

First, *Edge* is people.

As the late artist James Lee Byars and I once wrote: "To accomplish the extraordinary, you must seek extraordinary people." At the center of every *Edge* publication and event are remarkable people and remarkable minds. *Edge*, at its core, consists of the scientists, artists, philosophers, technologists, and entrepreneurs who are at the center of today's intellectual, technological, and scientific landscape.

Second, *Edge* is events. Through its special lectures, Master Classes, and annual dinners in California, London, Paris, and New York, *Edge* gathers together the "third-culture" scientific intellectuals and technology pioneers who are exploring the themes of the postindustrial age. In this regard, commenting about the 2008 *Edge* Master Class, "A Short Course in Behavioral Economics," science historian George Dyson wrote:

Retreating to the luxury of Sonoma to discuss economic theory in mid-2008 conveys images of fiddling while Rome burns. Do the architects of Microsoft, Amazon, Google, PayPal, and Facebook have anything to teach the behavioral economists—and anything to learn? So what? What's new? As it turns out, all kinds of things are new. Entirely new economic structures and pathways have come into existence in the past few years.

It was a remarkable gathering of outstanding minds. These are the people that are rewriting our global culture.

Third, *Edge* is a conversation.

Edge is different from the Algonquin Round Table or Bloomsbury Group, but it offers the same quality of intellectual adventure. Closer resemblances are the early seventeenth-century Invisible College, a precursor to the Royal Society. The Invisible College counted scientists such as Robert Boyle, John Wallis, and Robert Hooke among its members. Its common theme was to acquire knowledge through experimental investigation. Another inspiration is the Lunar Society of Birmingham, an informal club of the leading cultural figures of the new industrial age: James Watt, Erasmus Darwin, Josiah Wedgwood, Joseph Priestley, and Benjamin Franklin.

The online salon at Edge.org is a living document of millions of words that charts the *Edge* conversation over the past fifteen years. It is available, gratis, to the general public.

Edge.org was launched in 1996 as the online version of "The Reality Club," an informal gathering of intellectuals that met from 1981 to 1996 in Chinese restaurants, artist lofts, the boardrooms of Rockefeller University, the New York Academy of Sciences, investment banking firms, ballrooms, museums, living rooms, and elsewhere. Though the venue is now in cyberspace, the spirit of the Reality Club lives on in the lively back-and-forth discussions on the hot-button ideas driving the discussion today.

In the words of the novelist Ian McEwan, Edge.org is "open-minded, free-ranging, intellectually playful . . . an unadorned pleasure in curiosity, a collective expression of wonder at the living and inanimate world . . . an ongoing and thrilling colloquium."

In this, the second volume of the Best of *Edge* series, the focus is on ideas about "culture." We are pleased to present seventeen pieces, original works from the online pages of Edge.org, comprising edited interviews, commissioned essays, and transcribed talks, many of which are accompanied online with streaming video. While there's no doubt about the value of online presentations, books, whether bound and printed or presented electronically, are still an invaluable way to present important ideas. Thus, we are pleased to be able to offer this series of books to the public.

For this second volume, cutting-edge artists, inventors, mathematicians, computer scientists, visionaries, philosophers, and futurists explore new way of thinking about "culture."

Confronted with the truism that culture evolves, Tufts philosopher and cognitive scientist **Daniel C. Dennett** uses the evolution of music to illustrate "the way the traditional perspective on culture and the evolutionary perspective can join forces" in "The Evolution of Culture" (1999).

In "Why Do Some Societies Make Disastrous Decisions?" (2003) **Jared Diamond**, UCLA professor of geography and former professor of physiology, suggests "a road map of factors in failures of group decision-making." He divides the answers in four categories:

First of all, a group may fail to anticipate a problem before the problem actually arrives. Second, when the problem arrives, the group may fail to perceive the problem. Then, after they perceive the problem, they may fail even to try to solve the problem. Finally, they may try to solve it but may fail in their attempts to do so. While all this talking about reasons for failure and collapses of society may seem pessimistic, the flip side is optimistic: namely, successful decision-making. Perhaps if we understand the reasons why groups make bad decisions, we can use that knowledge as a checklist to help groups make good decisions.

The late **Denis Dutton**, philosopher and founder of *Arts & Letters Daily*, argues for a Darwinian explanation for human personality. In "Art and Human Reality" (2009), he defends Darwinian aesthetics, claiming it "is not some kind of ironclad doctrine that is supposed to replace a heavy post-structuralism with something just as oppressive. What surprises me about the resistance to the application of Darwin to psychology is the vociferous way in which people want to dismiss it, not even to consider it."

Musician and innovator **Brian Eno** answers his own question—"What is cultural value and how does that come about?"—in "A Big Theory of Culture" (1997). He proposes that

> Nearly all of art history is about trying to identify the source of value in cultural objects. Color theories and dimension theories, golden means, all those sort of ideas, assume that some objects are intrinsically more beautiful and meaningful than others. New cultural thinking isn't like that. It says that *we* confer value on things. *We* create the value in things. It's the act of conferring that makes things valuable. Now this is very important, because so many, in fact all fundamentalist ideas, rest on the assumption that some things have intrinsic value and resonance and meaning. All pragmatists work from another assumption: No, it's us. It's us who make those meanings.

In "We Are as Gods and Have to Get Good at It" (2009), environmentalist and visionary **Stewart Brand** notes that

Necessity comes from climate change, potentially disastrous for civilization. The planet will be okay, life will be okay. We will lose vast quantities of species, probably lose the rain forests if the climate keeps heating up. So it's a global issue, a global phenomenon. It doesn't happen in just one area. The planetary perspective now is not just aesthetic. It's not just perspective. It's actually a world-sized problem that will take world-sized solutions that involve forms of governance we don't have yet. It involves technologies we are just glimpsing. It involves what ecologists call ecosystem engineering. Beavers do it, earthworms do it. They don't usually do it on a planetary scale. We have to do it on a planetary scale. A lot of sentiments and aesthetics of the environmental movement stand in the way of that.

In "Turing's Cathedral" (2005), science historian **George Dyson** describes his visit to the Google headquarters: "I felt I was entering a 14th-century cathedral—not in the 14th century but in the 12th century, while it was being built." Dyson recalls H. G. Wells's 1938 prophecy—"The whole human memory can be, and probably in a short time will be, made accessible to every individual"—and argues that "Wells foresaw not only the distributed intelligence of the World Wide Web, but the inevitability that this intelligence would coalesce, and that power, as well as knowledge, would fall under its domain."

Computer scientist **David Gelernter** writes in "Time to Start Taking the Internet Seriously" (2010) that

The Internet is no topic like cell phones or video-game platforms or artificial intelligence; it's a topic like education. It's that big. Therefore beware: To become a teacher, master some topic you can teach; don't go to education school and master nothing. To work on the Internet, master some part of the Internet: engineering, software, computer science, communication theory, economics or business, literature or design. Don't go to Internet school and master nothing. There are brilliant, admirable people at Internet institutes. But if these institutes have the same effect on the Internet that education schools have had on education, they will be a disaster.

Mathematician **Karl Sigmund** explains in "Indirect Reciprocity, Assessment Hardwiring, and Reputation" (2004) that

Right now it turns out that economists are excited about this idea in the context of e-trading and e-commerce. In this case, you also have a lot of anonymous interactions, not between the same two people but within a hugely mixed group where you are unlikely ever to meet the same person again. Here the question of trusting the other, the idea of reputation, is particularly important. Google page rankings, the reputations of eBay buyers and sellers, and the Amazon.com reader reviews are all based on trust, and there is a lot of moral hazard inherent in these interactions.

Computer scientists and digital visionary, **Jaron Lanier** in "Digital Maoism: The Hazards of the New Online Collectivism" (2006) warns against the dangers of "the appeal of a new online collectivism." He argues in this essay that websites like Wikipedia are

nothing less than a resurgence of the idea that the collective is all-wise, that it is desirable to have influence concentrated in a bottleneck that can channel the collective with the most verity and force. This is different from representative democracy, or meritocracy. This idea has had dreadful consequences when thrust upon us from the extreme Right or the extreme Left in various historical periods. The fact that it's now being reintroduced today by prominent technologists and futurists, people whom in many cases I know and like, doesn't make it any less dangerous.

Clay Shirky, social software visionary and researcher, refutes Lanier in "On Jaron Lanier's 'Digital Maoism': An *Edge* Conversation" (2008) by stating:

> Curiously, the ability of the real Wikipedia to adapt to new challenges is taken at face value. The criticism is then directed instead at people proclaiming Wikipedia as an avatar of a golden era of collective consciousness. Let us stipulate that people who use terms like "hive mind" to discuss Wikipedia and other social software are credulous at best, and that their pronouncements tend towards caricature. What "Digital Maoism" misses is that Wikipedia doesn't work the way those people say it does.

Harvard physician and sociologist **Nicholas A. Christakis** investigates what causes networks to form and how networks operate in "Social Networks Are Like the Eye" (2008). He explains:

> The amazing thing about *social* networks, unlike other networks that are almost as interesting—networks of neurons or genes or stars or computers or all kinds of other things one can imagine—is that the nodes of a social network—the entities, the components—are themselves sentient, acting individuals who can respond to the network and actually form it themselves.

In "The Next Renaissance" (2008), social theorist **Douglas Rushkoff** argues past the claim that the Internet has offered us "personal" democracy via writing, to assert:

Writing is not the capability being offered us by these tools at all. The capability is programming—which almost none of us really know how to do. We simply use the programs that have been made for us, and enter our blog text in the appropriate box on the screen. Nothing against the strides made by citizen bloggers and journalists, but big deal. Let them eat blog.

In "Digital Power and Its Discontents" (2010), **Evgeny Morozov**, political commentator and blogger, and **Clay Shirky** debate on the subjects of dictators, democracy, Twitter revolutionaries, and the role of the Internet and social software in political lives of people living under authoritarian regimes. "You very quickly get a kind of philosophic vertigo," says Shirky. "You think you're asking a question about Twitter, and suddenly you realize you're asking a question about, say, Hayek."

W. Brian Arthur, pioneer in the field of new science of complexity and the economics of high technology, argues in "Does Technology Evolve?" (2009) that

> the two legs of the Theory of Evolution that are in technology are not at all Darwinian. They are quite different. They are (1) that certain existing building blocks are combined and recombined, and (2) that every so often some of those technologies get used to capture novel, newly discovered phenomena, which are in turn encapsulated as further building blocks. Most new technologies that come into being are only useful for their own purpose and don't form other building blocks. But occasionally, some do.

W. Daniel Hillis, the innovator and physicist who created the massively parallel computer, proposed in "*Aristotle*: The Knowledge Web" (2004):

With the knowledge web, humanity's accumulated store of information will become more accessible, more manageable, and more useful. Anyone who wants to learn will be able to find the best and the most meaningful explanations of what they want to know. Anyone with something to teach will have a way to reach those who want to learn. Teachers will move beyond their present role as dispensers of information and become guides, mentors, facilitators, and authors. The knowledge web will make us all smarter. The knowledge web is an idea whose time has come.

Richard Foreman, avant-garde playwright and director, presented *Edge* with a statement and a question in "The Pancake People" (2005). The statement appears in his program for his play *The Gods Are Pounding My Head*. The question is an opening to the future. Historian among futurists **George Dyson** responded with another question in "The Gödel-to-Google Net" (2005):

As Richard Foreman so beautifully describes it, we've been pounded into instantly available pancakes, becoming the unpredictable but statistically critical synapses in the whole Gödel-to-Google net. Does the resulting mind (as [Lewis Fry] Richardson would have it) belong to us? Or does it belong to something else?

In "The Age of the Informavore" (2009), **Frank Schirrmacher**, a prominent German author, journalist, and intellectual, notes:

We are apparently now in a situation where modern technology is changing the way people behave, people talk, people react, people think, and people remember. And you encounter this not only in a theoretical way, but when you meet people, when suddenly people start forgetting things, when suddenly people depend on their gadgets and other stuff to remember certain things. This is the beginning; it's just an experience. But if you think about it and you think about your own behavior, you suddenly realize that something fundamental is going on. There is one comment on *Edge* which I love which is in Daniel Dennett's response to the 2007 annual question: He writes that we have a population explosion of ideas, but not enough brains to cover them.

Culture

1.

The Evolution of Culture

Daniel C. Dennett

Philosopher; University Professor and codirector of the Center for Cognitive Studies at Tufts University; author, Breaking the Spell: Religion as a Natural Phenomenon

Cultures evolve. In one sense, this is a truism; in other senses, it asserts one or another controversial, speculative, unconfirmed theory of culture. Consider a cultural inventory of some culture at some time—say A.D. 1900. It should include all the languages, practices, ceremonies, edifices, methods, tools, myths, music, art, and so forth that compose that culture. Over time, that inventory changes. Today, a hundred years later, some items will have disappeared, some multiplied, some merged, some changed, and many new elements will appear for the first time. A verbatim record of this changing inventory through history would not be science; it would be a database. That is the truism: Cultures evolve over time. Everybody agrees about that. Now let's turn to the controversial question: How are we to explain the patterns to be found in that database? Are there any good theories or models of cultural evolution?

1. Science or Narrative?

One possibility is that the only patterns to be found in cultural evolution defy scientific explanation. They are, some might want to say, *narrative* patterns, not scientific patterns. There is clearly something to this, but it won't do as it stands, for many scientific patterns are also historical patterns, and hence are revealed and explained in narratives—of sorts. Cosmology, geology, and biology are all historical sciences. The great biologist D'Arcy Thompson once said: "Everything is the way it is because it got that way." If he is right—if *everything* is the way it is because it got that way—then every science must be, in part, a historical science. But not all history—all recounting of events in temporal sequence—is *narrative*, some might want to say. Human history is unique in that the patterns it exhibits require a different *form* of understanding: *hermeneutical* understanding, or *Verstehen*, or—you can count on the Germans to have lots of words for claims like this—*Geisteswissenschaft* (approximately: spiritual science). I think this too is partly right; there *is* a particular sort of understanding that is used to make sense of narratives about human agents. It is also true that the mark of a *good* story is that its episodes unfold *not* as the predicted consequences of general laws and initial conditions, but in delightfully surprising ways. These important facts do not show, however, that cultural evolution escapes the clutches of science and must be addressed in some other realm of inquiry. Quite the contrary: The humanistic comprehension of narratives and the scientific explanation of life processes, for all their differences of style and emphasis, have the same logical backbone. We can see this by examining the special form of understanding we use when following—and creating—good narratives.

Mediocre narratives are either a pointless series of episodes in temporal order—just "one damn thing after another"—or else so utterly predictable as to be boring. Between randomness and routine lie the good stories, whose surprising moments make sense in retrospect, in the framework provided by the unsurprising mo-

Daniel C. Dennett

ments. The perspective from which we can understand these narratives is what I have called *the intentional stance*: the strategy of analyzing the flux of events into *agents* and their (rational) *actions* and *reactions*. Such agents—people, in this case—*do things for reasons*, which can be predicted—up to a point—by cataloguing their reasons, their beliefs and desires, and calculating what, given those reasons, the most rational course of action for each agent would be. Sometimes the most rational course is flat obvious, so while the narrative is predictive (or true), it is uninteresting and unenlightening. To take a usefully simple case, a particular game of chess is interesting to the extent that we are surprised by either the brilliant moves that outstrip our own calculations of what it would be rational to do, or the blunders, which we thought too suboptimal to predict.

In the wider world of human activity, the same holds true. We don't find the tale of Jane going to the supermarket on her way home from work interesting precisely because it all unfolds so predictably from the intentional stance; today she never encountered any interesting options, given her circumstances. Other times, however, the most rational thing for an agent to do is far from obvious, and may be practically incalculable. When we encounter these narratives, we are surprised (and sometimes delighted, sometimes appalled) by the actual outcome. It makes sense in retrospect, but who'd have guessed that she'd decide to do *that*? The vast mass of routinely rational human behavior doesn't make good novels, but it is just such humdrum rational narrative that provides the background pattern that permits us to make sense, retrospectively, of the intriguing vagaries we encounter, and to anticipate the complications that will arise when the trains of events they put in motion collide.

The traditional model used by historians and anthropologists to try to explain cultural evolution uses the intentional stance as its explanatory framework. These theorists treat culture as being composed of goods, possessions of the people, who husband them in various ways, wisely or foolishly. People carefully preserve their traditions of fire-lighting, house-building, speaking, counting, justice, etc. They trade cultural items as they trade other goods. And of course some cultural items (wagons, pasta, recipes for chocolate cake, etc.) are definitely goods, and so we can plot their trajectories using the tools of economics. It is clear from this perspective that highly prized cultural entities will be protected at the expense of less favored cultural entities, and there will be a competitive market where agents both "buy" and "sell" cultural wares. If a new method of house-building or farming or a new style of music sweeps through the culture, it will be because people perceive advantages to these novelties.

The people, in this model, are seen as having an autonomous rationality: Deprive a person of his goods, and he stands there, naked but rational and full of informed desires. When he clothes himself and arms himself and equips himself with goods, he increases his powers, complicates his desires. If Coca-Cola bottles proliferate around the world, it is because more and more people prefer to buy a Coke. Advertising may fool them. But then we look to the advertisers, or those who have hired them, to find the relevant agents whose desires fix the values for our cost-benefit calculations. *Cui bono?* Who benefits? The purveyors of the goods, and those they hire to help them, etc. In this way of thinking, then, the relative "replicative" power of various cultural goods—whether Coke bottles, building styles, or religious creeds—is measured in the marketplace of cost-benefit calculations performed by the people.

Daniel C. Dennett

Biologists, too, can often make sense of the evolution (in the neutral sense) of features of the natural world by treating them as goods belonging to various members of various species: one's food, one's nest, one's burrow, one's territory, one's mate[s], one's time and energy. Cost-benefit analyses shed light on the husbandry engaged in by the members of the different species inhabiting some shared environment.* Not every "possession" is considered a good, however. The dirt and grime that accumulate on one's body, to say nothing of the accompanying flies and fleas, are of no value, or of negative value, for instance. These hitchhikers are not normally considered as goods by biologists, except when the benefits derived from them (by *whom?*) are manifest.

This traditional perspective can obviously explain many features of cultural and biological evolution, but it is not uniformly illuminating, nor is it obligatory. I want to show how theorists of culture—historians, anthropologists, economists, psychologists, and others—can benefit from adopting a different vantage point on these phenomena. It is a different application of the intentional stance, one which still quite properly gives pride of place to the *Cui bono* question, but which can provide alternative answers that are often overlooked. The perspective I am talking about is Richard Dawkins's *meme's-eye point of view*, which recognizes—and takes seriously—the possibility that cultural entities may evolve according to selectional regimes that make sense only when the answer

---

* Such organisms need not be deemed to be making conscious decisions, of course, but the rationality, such as it is, of the "decisions" they make is typically anchored to the expected benefit to the individual organism. See Elliott Sober and David Sloan Wilson's *Unto Others: The Evolution and Psychology of Unselfish Behavior* (1998) for important discussions of gene, individual, and group benefits of such decision-making.

to the *Cui bono* question is that it is the cultural items *themselves* that benefit from the adaptations they exhibit.*

2. Memes as Cultural Viruses

Whenever costs and benefits are the issue, we need to ask, *Cui bono?* A benefit by itself is not explanatory; a benefit in a vacuum is indeed a sort of mystery. Until it can be shown how the benefit actually redounds to enhance the replicative power of a replicator, it just sits there, alluring, perhaps, but incapable of explaining anything.

We see an ant laboriously climbing up a stalk of grass. Why is it doing that? Why is that adaptive? What good accrues to the ant by doing that? That is the wrong question to ask. No good at all accrues to the ant. Is it just a fluke, then? In fact, that's exactly what it is: a fluke! Its brain has been invaded by a fluke (*Dicrocoelium dendriticum*), one of a gang of tiny parasites that need to get them-

---

* Sober and Wilson (1998) note that there is a gap in their model of cultural evolution: "We can say that functionless [relative to human individual and group fitness] behavior should be more common in humans than other species, but we cannot explain why a particular functionless behavior has evolved in a particular culture. That kind of understanding probably requires detailed historical knowledge of the culture, and it may turn out that some behaviors evolved mainly by chance" (p. 171). Dawkins's theory of memes, as briefly sketched in a single chapter of *The Selfish Gene* (1976, but see also Dawkins, 1993), is hardly a theory at all, especially compared to the models of cultural evolution developed by other biologists, such as Cavalli-Sforza and Feldman (1981), Lumsden and Wilson (1981), and Boyd and Richerson (1985). Unlike these others, Dawkins offers no formal development, no mathematical models, no quantitative predictions, no systematic survey of relevant empirical findings. But Dawkins does present an idea that is overlooked by all the others, including Sober and Wilson, and it is, I think, a most important idea. It is the key to understanding how we can be not just guardians and transmitters of culture, but cultural entities ourselves—all the way in.

Daniel C. Dennett

selves into the intestines of a sheep in order to reproduce (Ridley, 1995, p. 258). (Salmon swim upstream; these parasitic worms drive ants up grass stalks to improve their chances of being ingested by a passing sheep.) The benefit is not to the reproductive prospects of the ant but to the reproductive prospects of the fluke.*

Dawkins points out that we can think of cultural items, memes, as parasites, too. Actually, they are more like a simple virus than a worm. Memes are supposed to be analogous to genes, the replicating entities of the cultural media, but they also have vehicles, or phenotypes; they are like not-so-naked genes. They are like viruses (Dawkins, 1993). Basically, a virus is just a string of nucleic acid with attitude—and a protein overcoat. A viroid is an even more naked gene. And similarly, a meme is an information packet with attitude—with some phenotypic clothing that has differential effects in the world that thereby influence its chances of getting replicated. (What is a meme *made of*? It is made of information, which can be carried in *any* physical medium. More on this later.)

And in the domain of memes, the ultimate beneficiary, the beneficiary in terms of which the final cost-benefit calculations must apply, is the meme itself, not its carriers. This is not to be heard as a bold empirical claim, ruling out (for instance) the role of individual human agents in devising, appreciating, and securing the spread and prolongation of cultural items. As I have already noted, the traditional perspective on cultural evolution handsomely explains many of the patterns to be observed. My proposal is rather

---

* Strictly speaking, this is to benefit the reproductive prospects of the fluke's genes (or the genes of the fluke's "group"), for as Sober and Wilson (1998) point out (p. 18) in their use of *D. dendriticum* as an example of altruistic behavior, the fluke who actually does the driving in the brain is a sort of kamikaze pilot, dying without any chance of passing on its own genes while benefiting its (asexually reproduced) near-clones in other parts of the ant.

that we adopt a perspective or point of view from which a wide variety of different empirical claims can be compared, *including the traditional claims*, and the evidence for them considered in a neutral setting, a setting that does not prejudge these hot-button questions.

In the analogy with the fluke, we are invited to consider a meme to be like a parasite that commandeers an organism for its own replicative benefit, but we should remember that such hitchhikers or *symbionts* can be classified into three fundamental categories:

- *parasites*, whose presence lowers the fitness of their host;
- *commensals*, whose presence is neutral (though, as the etymology reminds us, they "share the same table"); and
- *mutualists*, whose presence enhances the fitness of both host and guest.

Since these varieties are arrayed along a continuum, the boundaries between them need not be too finely drawn; just where benefit drops to zero or turns to harm is not something to be directly measured by any practical test, though we can explore the consequences of these turning points in models.

We should expect memes to come in all three varieties, too. This means, for instance, that it is a mistake to assume that the "cultural selection" of a cultural trait is always "for cause"—always because of some perceived (or even misperceived) benefit it provides to the host. We can always ask if the hosts, the human agents who are the *vectors*, perceive some benefit and (for that reason, good or bad) assist in the preservation and replication of the cultural item in question, but we must be prepared to entertain the answer that they do not. In other words, we must consider as a real possibility the hypothesis that the human hosts are, individually

Daniel C. Dennett

or as a group, either oblivious to, or agnostic about, or even positively dead set against, some cultural item, which nevertheless is able to exploit its hosts as vectors.

The most familiar cases of cultural transmission and evolution—the cases that tend to be in the spotlight—are innovations that are obviously of some direct or indirect benefit to the genetic fitness of the host. A better fishhook catches more fish, feeds more bellies, makes for more surviving grandchildren, etc. The only difference between stronger arms and a better fishhook in the (imagined) calculation of impact on fitness is that the stronger arms might be passed on quite directly through the germ line, while the fishhook definitely must be culturally transmitted. (The stronger arms *could* be culturally transmitted as well. A tradition of bodybuilding, for instance, could explain why there was very low [genetic] heritability for strong adult arms, and yet a very high rate of strong adult arms in a population.) But however it might be that strong arms or fishhooks are transmitted, they are typically supposed to be a good bargain from the perspective of genetic fitness. The bargain might, however, be myopic—only good in the short run. After all, even agriculture, in the long run, may be a dubious bargain if what you are taking as your *summum bonum* is Darwinian fitness (see Jared Diamond, *Guns, Germs, and Steel,* 1997, for fascinating reflections on the uncertain benefits of abandoning the hunter-gatherer lifestyle). What alternatives are there?

First, we need to note that in the short run (evolutionarily speaking—that is, from the perspective of a few centuries or even millennia) something might flourish in a culture independent of whether it was of actual benefit to genetic fitness, but strongly linked to whether it was of *apparent* benefit to genetic fitness. Even if you think that Darwinian fitness enhancement is the principle driving engine of cultural evolution, you have to posit some swifter, more immediate mechanism of retention and transmission. It's not hard to find one. We are genetically endowed with a biased quality space: Some things feel good and some things don't. We tend to live by the rule *If it feels good, keep it.* This rough and ready rule can be tricked, of course. The sweet tooth is a standard example. The explosion of cultural items—artifacts, practices, recipes, patterns of agriculture, trade routes—that depend quite directly on the exploitation of the sweet tooth has probably had a considerable net *negative* effect on human genetic fitness. Notice that explaining the emergence of these cultural items by citing their "apparent" benefit to genetic fitness does not in any way commit us to the claim that people think they are enhancing their genetic fitness by acquiring and consuming sugar. The rationale is not theirs, but Mother Nature's. They just go with what they like.

Still, given what people innately like, they go on to figure out, ingeniously and often with impressive foresight, how to obtain what they like. This is still the traditional model of cultural evolution, with people husbanding their goods in order to maximize what they prefer—and getting their preferences quite directly from their genetic heritage. But this very process of rational calculation can lead to more interesting possibilities. As such an agent complicates her life, she will almost certainly acquire new preferences that are themselves culturally transmitted symbionts of one sort or another. Her sweet tooth may lead her to buy a cookbook, which inspires her to enroll in a culinary arts program, which turns out to be so poorly organized that she starts a student protest movement, in which she is so successful that she is invited to head an educational reform movement, for which a law degree would be a useful credential, and so on. Each new goal will have to bootstrap itself into the memosphere by exploiting some preestablished preference, but this recursive process, which can proceed at

Daniel C. Dennett

breakneck speed relative to the glacial pace of genetic evolution, can transform human agents indefinitely far away from their genetic beginnings. In an oft-quoted passage, E. O. Wilson claimed otherwise: "The genes hold culture on a leash. The leash is very long, but inevitably values will be constrained in accordance with their effects on the human gene pool" (Wilson, 1978, p. 167).

But Wilson's leash is indefinitely long and elastic. Consider the huge space of *imaginable* cultural entities, practices, values. Is there any point in that vast space that is utterly unreachable? Not that I can see. The constraints Wilson speaks of can be so co-opted, exploited, and blunted in a recursive cascade of cultural products and meta-products that there may well be traversable paths to every point in that space of imaginable possibilities. I am suggesting, that is, that cultural possibility is less constrained than genetic possibility. We can articulate persuasive biological arguments to the effect that certain imaginable species are unlikely in the extreme—flying horses, unicorns, talking trees, carnivorous cows, spiders the size of whales—but neither Wilson nor anybody else to my knowledge has yet offered parallel grounds for believing that there are similar obstacles to trajectories in imaginable cultural design space. Many of these imaginable points in design space would no doubt be genetic cul-de-sacs, in the sense that any lineage of *H. sapiens* that ever occupied them would eventually go extinct as a result, but this dire prospect is no barrier to the evolution and adoption of such memes in the swift time of cultural history.* To combat Wilson's metaphor with one of my own: The

---

* Boyd and Richerson (1992) show that "virtually any behavior can become stable within a social group if it is sufficiently buttressed by social norms" (Sober and Wilson, 1998, p. 152). Our biology strongly biases us to value health, nutritious food, the avoidance of bodily injury, and of course having lots of offspring, so a sheltered

genes provide not a leash but a launching pad, from which you can get almost anywhere, by one devious route or another. It is precisely in order to explain the patterns in cultural evolution that are *not* strongly constrained by genetic forces that we need the memetic approach.

The memes that proliferate will be the memes that replicate one way or another—by hook or by crook. Think of them as entering the brains of culture members, making phenotypic alterations thereupon, and then submitting themselves to the great selection tournament—not the Darwinian genetic fitness tournament (life is too short for that) but the Dawkinsian meme-fitness tournament. It is their fitness as memes that is on the line, not their host's genetic fitness. And the environments that embody the selective pressures that determine their fitness are composed in large measure of other memes.

Why do their hosts put up with this? Why should the overhead costs of establishing a whole new system of differential reproduction be borne by members of *H. sapiens*? Note that the question to be asked and answered here is parallel to the question we ask about any symbiont-host relationship: Why do the hosts put up with it? And the short answer is that it is too costly to eradicate, but this just means that the benefits accruing to the machinery that is being exploited by the parasites are so great that keeping the machinery and tolerating the parasites (to the extent that they are tolerated) has so far been the best deal available. And whether or not in the long run (millions of years) this infestation will be

theorist might suppose that it is highly unlikely that any human group could ever support a fashion for, say, bodily fragility or bulimia, or the piercing of bodily parts, or suicide, or celibacy. If even these practices can so readily overturn our innate biases, where can Wilson's leash do any serious constraining?

Daniel C. Dennett

viewed as mutualism or commensalism or parasitism, in the short run (the previous few millennia) the results have been spectacular: the creation of a new biological type of entity: a person.

I like to compare this development to the revolution that happened among the bacteria roughly a billion years ago. Relatively simple prokaryotes got invaded by some of their neighbors. The resulting endosymbiotic teams were more fit than their uninfected cousins, and they prospered. These eukaryotes, living alongside their prokaryotic cousins, but enormously more complex, versatile, and competent thanks to their hitchhikers, opened up the design space of multicellular organisms. Similarly, the emergence of culture-infected hominids has opened up yet another region of hitherto unoccupied and untraversable design space. We live alongside our animal cousins, but we are enormously more complex, versatile, and competent. Our brains are bigger, to be sure, but it is mainly due to their infestation by memes that they gain their powers. Joining forces with our own memes, we create new candidates for the locus of benefit, new answers to *Cui bono?*

3. Darwin's Path to Memetic Engineering

The meme's-eye view doesn't just open up new vistas for the understanding of patterns in culture; it also provides the foundation for answering a question left dangling by the traditional model of cultural evolution. The traditional view presupposes rational self-interested agents, intent on buying and selling and improving their lot. *Where did they come from?* The standard background assumption is that they are just animals, whose *Cui bono* question is to be dealt with in terms of its impact on genetic fitness, as we have seen. But when people acquire other interests, including interests directly opposed to their genetic interests, they enter a new space of possibilities—something no salmon or fruit fly or bear can do. How could this great river of novelty get started?

Here I think we can get help from Darwin's opening exposition of the theory of natural selection. In the first chapter of *Origin of Species*, Darwin introduces his great idea of natural selection by an ingenious expository device, an instance of the very gradualism that he was about to discuss. He begins not with natural selection—his destination—but what he calls *methodical selection*: the deliberate, foresighted, intended "improvement of the breed" by animal and plant breeders. He begins, in short, with familiar and uncontroversial ground that he can expect his readers to share with him.

> We cannot suppose that all the breeds were suddenly produced as perfect and as useful as we now see them; indeed, in several cases, we know that this has not been their history. The key is man's power of accumulative selection: nature gives successive variations; man adds them up in certain directions useful to him (p. 30, Harvard facsimile edition).

But, he goes on to note, in addition to such methodical selection, there is another process, which lacks foresight and intention and which he calls *unconscious selection*:

> At the present time, eminent breeders try by methodical selection, with a distinct object in view, to make a new strain or sub-breed, superior to anything existing in the country. But, for our purpose, a kind of Selection, which may be called Unconscious, and which results from every one trying to possess and breed from the best individual animals, is more important. Thus, a man who intends keeping pointers naturally tries to get as good dogs as he can, and

Daniel C. Dennett

afterwards breeds from his own best dogs, but he has no wish or
expectation of permanently altering the breed (p. 34).

Long before there was deliberate breeding, unconscious selection was the process that created and refined all our domesticated
species, and even at the present time, unconscious selection continues. Darwin gives a famous example:

> There is reason to believe that King Charles's spaniel has been
> unconsciously modified to a large extent since the time of that
> monarch (p. 35).

There is no doubt that unconscious selection has been a major
force in the evolution of domesticated species. On unconscious
selection of both domesticated plants and animals, see Diamond
(1997). In our own time, unconscious selection goes on apace, and
we ignore it at our peril. Unconscious selection in bacteria and viruses for resistance to antibiotics is only the most notorious and
important example. Consider the "genes for longevity" that have
recently been bred into laboratory animals such as mice and rats.
It is probably true, however, that much if not all of the effect that
has been obtained in these laboratory breeding experiments has
simply undone the unconscious selection for short-livedness at
the hands of the suppliers of those laboratory animals. The stock
the experimenters started with had shorter life expectancy than
their wild cousins simply because they had been bred for many generations for early reproductive maturity and robustness, and short
lives came along as an unintended (unconscious) side consequence
(Daniel Promislow, personal correspondence).
Darwin pointed out that the line between unconscious and methodical selection was itself a fuzzy, gradual boundary:

The man who first selected a pigeon with a slightly larger tail, never dreamed what the descendants of that pigeon would become through long-continued, partly unconscious and partly methodical selection (p. 39).

And both unconscious and methodical selection, he notes finally, are but special cases of an even more inclusive process, natural selection, in which the role of human intelligence and choice stands at zero. From the perspective of natural selection, changes in lineages due to unconscious or methodical selection are merely changes in which one of the most prominent selective pressures in the environment is human activity. It is not restricted, as we have seen, to domesticated species. White-tailed deer in New England now seldom exhibit the "white flag" of a bobbing tail during head-long flight that was famously observed by early hunters; the arrival of human beings today is much more likely to provoke them to hide silently in underbrush than to flee. Those white flags were too easy a target for hunters with guns, it seems.

This nesting of different processes of natural selection now has a new member: genetic engineering. How does it differ from the methodical selection of Darwin's day? It is just less dependent on the preexisting variation in the gene pool and proceeds more directly to new candidate genomes, with less overt and time-consuming trial and error. Darwin had noted that in his day, "Man can hardly select, or only with much difficulty, any deviation of structure excepting such as is externally visible; and indeed he rarely cares for what is internal," but today's genetic engineers have carried their insight into the molecular innards of the organisms they are trying to create. There is ever more accurate foresight, but even here, if we look closely at the practices in the

Daniel C. Dennett

laboratory, we will find a large measure of exploratory trial and error in the search of the best combinations of genes.

We can use Darwin's three levels of genetic selection, plus our own fourth level, genetic engineering, as a model for four parallel levels of *memetic* selection in human culture. In a speculative spirit, I am going to sketch how it might go, using an example that has particularly challenged some Darwinians and hence been held up as a worthy stumbling block: a cultural treasure untouchable by evolutionists: music. Music is unique to our species, but found in every human culture. It is manifestly complex, intricately designed, an expensive consumer of time, energy, and materials. How did music start? What was or is the answer to its *Cui bono* question? Steven Pinker is one Darwinian who has declared himself baffled about the possible evolutionary origins and survival of music, but that is because he has been looking at music in the old-fashioned way, looking for it to have some contribution to make to the genetic fitness of those who make and participate in the proliferation of music.* There may well be some such effect that is important, but I want to make the case that there might also be a purely *memetic* explanation of the origin of music. Here, then, is my Just-so Story, working gradually up Darwin's hierarchy of kinds of selection.

* "What benefit could there be to diverting time and energy to the making of plinking noises, or to feeling sad when no one has died? . . . As far as biological cause and effect are concerned, music is useless" (Steven Pinker, *How the Mind Works*, 1997, p. 528). On p. 538, he contrasts music with the other topics of his book: "I chose them as topics because they show the clearest signs of being adaptations. I chose music because it shows the clearest signs of not being one."

Natural Selection of Musical Memes

One day one of our distant hominid ancestors sitting on a fallen log happened to start banging on it with a stick—*boom, boom, boom.* For *no good reason at all.* This was just idle diddling, a by-product, perhaps, of a slightly out-of-balance endocrine system. This was, you might say, mere nervous fidgeting, but the repetitive sounds striking his ears just happened to feel to him like a slight improvement on silence. A feedback loop was closed, and the *repetition— boom, boom, boom*—was "rewarding." If we leave this individual all by himself, drumming away on his log, then we would say that he had simply developed a habit, *possibly* therapeutic in that it "relieved anxiety," but just as possibly a *bad* habit—a habit that did him and his genes no good at all, but just exploited a wrinkle that happened to exist in his nervous system, creating a feedback loop that tended to lead to individual replications of drumming by him under various circumstances. No musical appreciation, no insight, no goal or ideal or project need be imputed to our solitary drummer.

Now introduce some other ancestors who happen to see and hear this drummer. They might pay no attention, or be irritated enough to make him stop or drive him away, or they might, again *for no reason*, find their imitator-circuits tickled into action; they might feel an urge to drum along with musical Adam. What are these imitator-circuits I've postulated? Just whatever it takes to make it somewhat more likely than not that some activities by conspecifics are imitated—a mere reflex, if you like, of which we may see a fossil trace when spectators at a football match cannot help making shadow-kicking motions more or less in unison with the players on the field. One can postulate reasons why having some such imitative talents built in would be a valuable adaptation—

Daniel C. Dennett

one that enhances one's genetic fitness—but while this is both plausible and widely accepted, it is strictly speaking unnecessary for my Just-so Story. The imitative urge might just as well be a functionless by-product of some other adaptive feature of the human nervous system. Suppose, then, that for no good reason at all, the drumming habit is *infectious*. When one hominid starts drumming, soon others start drumming along in imitation. This could happen. A perfectly pointless practice, of no utility or fitness-enhancing benefit at all, could become established in a community. It might be positively detrimental: The drumming scares away the food, or uses up lots of precious energy. It would then be just like a disease, spreading simply because it *could* spread, and lasting as long as it could find hosts to infect. If it were detrimental in this way, variant habits that were less detrimental—less virulent—would tend to evolve to replace it, other things being equal, for they would tend to find more available healthy hosts to migrate to. And of course such a habit *might* even provide a positive benefit to its hosts (enhancing their reproductive chances—a familiar dream of musicians everywhere, and it might be true, or have been true in the past). But providing a genetic benefit of this sort is only one of the paths such a habit might pursue in its mindless quest for immortality. Habits—good, bad, and indifferent—could persist and replicate, unappreciated and unrecognized, for an indefinite period of time, provided only that the replicative and dispersal machinery is provided for them. The drumming virus is born.

Let me pause to ask the question: What is such a habit made of? What gets passed from individual to individual when a habit is copied? Not stuff, not packets of material, but pure information, the information that generates the pattern of behavior that replicates. A cultural virus, unlike a biological virus, is not tethered to any particular physical medium of transmission.*

Unconscious Selection of Memes

On with our Just-so Story. Some of the drummers begin to hum, and of all the different hums, a few are more infectious than the rest, and those hominids who happen to start the humming in these ways become the focus of attention, as sources of humming. A competition between different humming patterns emerges. Here we can begin to see the gradual transition to unconscious selection. Suppose that being such a focus of humming happens to feel good—whether or not it enhances one's genetic fitness slightly (it might, of course; perhaps the females tend to be more receptive to those who start the winning hums). The same transition to unconscious selection can be seen among viruses and other pathogens, by the way. If scratching an itch feels good, and also has the side effect of keeping a ready supply of viral emigrés on one's fingertips, the part of the body most likely to come in contact with another host, one is unconsciously selecting for just such a mode of transmission by one's myopic and uncomprehending preference for scratching when one itches—and this does not depend on scratching having any fitness-enhancing benefits *for you*: It may

---

* This is not the decisive difference some critics of memes have declared it to be. We can readily enough imagine virus-like symbionts that have alternate transmission media—that are (roughly) indifferent to whether they arrive at new hosts by direct transportation (as with regular bacteria, viruses, viroids, fungi . . .) or by something akin to the messenger-RNA transcription process: They stay in their original hosts, but imprint their information on some messenger element (rather like a prion, we may imagine) that then is broadcast, only to get transcribed in the host into a copy of the "sender." And if there could be two such communication channels, there could be twelve or a hundred, just as there are for transmission of cultural habits.

Daniel C. Dennett

be, like the ant's hankering for the top of the grass stem, a desire that benefits the parasite, not the host. Similarly, if varying the tempo and pitch of one's hums feels good and also happens to create a ready supply of more attention-holding noises for spreading to conspecifics, one's primitive aesthetic preference can begin to shape, unconsciously, the lineages of humming habit that spread through one's community.

Brains in the community begin to be infected by a variety of these memes. Competition for time and space in these brains becomes more severe. The infected brains begin to take on a structure, as the memes that enter "learn" to cooperate on the task of turning a brain into a proper meme-nest, with lots of opportunities for entrance and exit (and hence replication).* Meanwhile, any memes out there "looking for" hosts will have to compete for available space therein. Just like germs.

Methodical Selection of Memes

As the structure grows, it begins to take on a more active role in selecting. That is to say, the brains of the hosts, like the brains of the owners of domesticated animals, become ever more potent and discerning selective agencies—still largely unwitting, but nevertheless having a powerful influence. Some people, it turns out, are better at this than others. As Darwin says of animal breeders, "Not one man in a thousand has accuracy of eye and judgment sufficient to become an eminent breeder."

We honor Bach, the artistic genius, but he was no "natural"

---

* Sober and Wilson (1998) describe circumstances in which individuals of unrelated lineages thrown into group situations can be selected for cooperativity. Just how—if at all—this model can be adapted for memetic coalescence is a topic for further research.

doodler, no mere intuitive genius just "playing by ear." He was the master musical technologist of his day, the inheritor of musical instruments whose designs had been honed over several millennia, as well as the beneficiary of some relatively recent additions to the music-maker's toolbox—a fine system of musical notation, keyboard instruments that permitted the musician to play many notes at once, and an explicit, codified, rationalized *theory* of counterpoint. These mind-tools were revolutionary in the way they opened up musical design space for Bach and his successors.

And Bach, like the one man in a thousand who has the discernment to be an eminent animal breeder, knew how to breed new strains of music from old. Consider, for instance, his hugely successful chorale cantatas. Bach shrewdly chose, for his breeding stock, *chorales*—hymn melodies that had already proven themselves to be robust inhabiters of their human hosts, *already domesticated* tunes his audiences had been humming for generations, building up associations and memories, memes that had already sunk their hooks deeply into the emotional habits and triggers of the brains where they had been replicating for years. Then he used his technology to create variations on these memes, seeking to strengthen their strengths and dampen their weaknesses, putting them in new environments, inducing new hybrids.

Memetic Engineering

What about memetic engineering? Was Bach, by virtue of his highly sophisticated approach to the design of replicable musical memes, not just a *meme-breeder* but a *memetic engineer*? In the light of Darwin's admiring comment on the rare skill—the genius—of the good breeder, it is interesting to note how sharply our prevailing attitudes distinguish between our honoring the "art" of selective breeding and our deep suspicion and disapproval of the

Daniel C. Dennett

"technology" of gene-splicing. Let's hear it for *art*, but not for *technology*, we say, forgetting that the words share a common ancestor, *techné*, the Greek word for art, skill, or craft in any work. We retreat in horror from genetically engineered tomatoes and turn up our noses at "artificial" fibers in our clothing, while extolling such "organic" and "natural" products as whole-grain flour or cotton and wool, forgetting that grains and cotton plants and sheep are themselves products of human technology, of skillful hybridization and rearing techniques. He who would clothe himself in fibers unimproved by technology and live on food from nondomesticated sources is going to be cold and hungry indeed.

Besides, just as genetic engineers, for all their foresight and insight into the innards of things, are still at the mercy of natural selection when it comes to the fate of their creations (that is why, after all, we are so cautious about letting them release their brainchildren on the outside world), so too the memetic engineer, no matter how sophisticated, still has to contend with the daunting task of winning the replication tournaments in the memosphere. One of the most sophisticated musical memetic engineers of the age, Leonard Bernstein, wryly noted this in a wonderful piece he entitled "Why Don't You Run Upstairs and Write a Nice Gershwin Tune?" (which appeared in the *Atlantic Monthly* in April 1955 and was reprinted in *The Joy of Music*, 1959, pp. 52–62).

Bernstein had credentials and academic honors aplenty in 1955, but no songs on the hit parade.

> A few weeks ago a serious composer-friend and I . . . got boiling mad
> about it. Why shouldn't we be able to come up with a hit, we said, if
> the standard is as low as it seems to be? We decided that all we had to
> do was to put ourselves into the mental state of an idiot and write a
> ridiculous hillbilly tune.

They failed—and not for lack of trying. As Bernstein wistfully remarked, "It's just that it would be nice to hear someone accidentally whistle something of mine, somewhere, just once."

His wish came true, of course, a few years later, when *West Side Story* burst into the memosphere.

4. Conclusions

There is surely much, much more to be said—to be discovered—about the evolution of music. I chose it as my topic because it so nicely illustrates the way the traditional perspective on culture and the evolutionary perspective can join forces, instead of being seen to be in irresolvable conflict. If you believe that music is *sui generis*, a wonderful, idiosyncratic feature of our species that we prize in spite of the fact that it has *not* been created to enhance our chances of having more offspring, you may well be right—*and if so, there is an evolutionary explanation of how this can be true*. You cannot evade the obligation to explain how such an expensive, time-consuming activity came to flourish in this cruel world, and a Darwinian theory of culture is an ally, not an opponent, in this investigation.

While it is true that Darwin wished to contrast the utter lack of foresight or intention in natural selection with the deliberate goal-seeking of the artificial or methodical selectors, in order to show how the natural process could in principle proceed without any mentality at all, he did not thereby establish (as many seem to have supposed) that deliberate, goal-directed, intentional selection is not a subvariety of natural selection! There is no conflict between the claim that artifacts (including abstract artifacts—memes) are the products of natural selection, and the claim that they are (often) the foreseen, designed products of intentional human activity.

Daniel C. Dennett

Some memes are like domesticated animals; they are prized for their benefits, and their replication is closely fostered and relatively well understood by their human owners. Some memes are more like rats; they thrive in the human environment in spite of being positively selected against—ineffectually—by their unwilling hosts. And some are more like bacteria or viruses, commandeering aspects of human behavior (provoking sneezing, for instance) in their "efforts" to propagate from host to host. There is artificial selection of "good" memes—like the memes of arithmetic and writing, the theory of counterpoint, and Bach's cantatas, which are carefully taught to each new generation. And there is unconscious selection of memes of all sorts—like the subtle mutations in pronunciation that spread through linguistic groups, presumably with some efficiency advantage, but perhaps just hitchhiking on some quirk of human preference. And there is unconscious selection of memes that are positively a menace, but which prey on flaws in the human decision-making apparatus, as provided for in the genome and enhanced and adjusted by other cultural innovations—such as the abducted-by-aliens meme, which makes perfect sense when its own fitness as a cultural replicator is considered. Only the meme's-eye perspective unites all these possibilities under one view.

Finally, one of the most persistent sources of discomfort about memes is the dread suspicion that an account of human minds in terms of brains being parasitized by memes will undermine the precious traditions of human creativity. On the contrary, I think it is clear that *only* an account of creativity in terms of memes has much of a chance of giving us any way to *identify with* the products of our own minds. We human beings extrude other products, on a daily basis, but after childhood, we don't tend to view our feces with the pride of an author or artist. These are mere biological by-products, and although they have their own modest individuality and idiosyncrasy, it is not anything we cherish. How could we justify viewing the secretions of our poor infected brains with any more pride? Because we *identify with* some subset of the memes we harbor. Why? Because among the memes we harbor are those that put a premium on identifying with just such a subset of memes! Lacking that meme-borne attitude, we would be mere *loci* of interaction, but we have such memes—that is who we are.

Daniel C. Dennett

2.
Why Do Some Societies Make Disastrous Decisions?

Jared Diamond

*Professor of Geography at the University of California–Los Angeles;
author,* Guns, Germs, and Steel *and* Collapse

Education is supposed to be about teachers imparting knowledge
to students. As every teacher knows, though, if you have a good
group of students, education is also about students imparting
knowledge to their supposed teachers and challenging their assumptions. That's an experience that I've been through in the last
couple of months, when for the first time in my academic career I
gave a course to undergraduates, highly motivated UCLA undergraduates, on collapses of societies. Why is it that some societies
in the past have collapsed while others have not? I was discussing famous collapses such as those of the Anasazi in the U.S.
Southwest, Classic Maya civilization in the Yucatán, Easter Island
society in the Pacific, Angkor Wat in southeast Asia, Great Zim-
babwe in Africa, Fertile Crescent societies, and Harappan Indus
Valley societies. These are all societies that we've realized, from
archaeological discoveries in the last twenty years, hammered
away at their own environments and destroyed themselves in part
by undermining the environmental resources on which they depended.

For example, the Easter Islanders, Polynesian people, settled an
island that was originally forested, and whose forests included the

world's largest palm tree. The Easter Islanders gradually chopped down that forest to use the wood for canoes, firewood, transporting statues, raising statues, and carving and also to protect against soil erosion. Eventually they chopped down all the forests to the point where all the tree species were extinct, which meant that they ran out of canoes, they could no longer erect statues, there were no longer trees to protect the topsoil against erosion, and their society collapsed in an epidemic of cannibalism that left 90 percent of the islanders dead. The question that most intrigued my UCLA students was one that hadn't registered on me: How on earth could a society make such an obviously disastrous decision as to cut down all the trees on which they depended? For example, my students wondered, what did the Easter Islanders say as they were cutting down the last palm tree? Were they saying, Think of our jobs as loggers, not these trees? Were they saying, Respect my private property rights? Surely the Easter Islanders, of all people, must have realized the consequences to themselves of destroying their own forest. It wasn't a subtle mistake. One wonders whether—if there are still people left alive a hundred years from now—people in the next century will be equally astonished about our blindness today as we are about the blindness of the Easter Islanders.

This question, why societies make disastrous decisions and destroy themselves, is one that not only surprised my UCLA undergraduates, but also astonishes professional historians studying collapses of past societies. The most cited book on the subject of the collapse of societies is by the historian Joseph Tainter. It's entitled *The Collapse of Complex Societies*. Tainter, in discussing ancient collapses, rejected the possibility that those collapses might be due to environmental management because it seemed so unlikely to him. Here's what Tainter said:

Jared Diamond

With their administrative structure, and capacity to allocate both labor and resources, dealing with adverse environmental conditions may be one of the things that complex societies do best. It is curious that they would collapse when faced with precisely those conditions that they are equipped to circumvent. . . .

As it becomes apparent to the members or administrators of a complex society that a resource base is deteriorating, it seems most reasonable to assume that some rational steps are taken toward a resolution.

Joseph Tainter concluded that the collapses of all these ancient societies couldn't possibly be due to environmental mismanagement, because they would never make these bad mistakes. Yet it's now clear that they did make these bad mistakes.

My UCLA undergraduates, and Tainter as well, have identified a very surprising question: namely, failures of group decision-making on the part of whole societies, or governments, or smaller groups, or businesses, or university academic departments. The question of the failure of group decision-making is similar to questions of failures of individual decision-making. Individuals make bad decisions: They enter bad marriages, they make bad investments, their businesses fail. But in failures of group decision-making, there are some additional factors, notably conflicts of interest among the members of the group, that don't arise with failures of individual decision-making. This is obviously a complex question; there's no single answer to it. There are no agreed-on answers.

What I'm going to suggest is a road map of factors in failures of group decision-making. I'll divide the answers into a sequence of four somewhat fuzzily delineated categories. First of all, a group may fail to anticipate a problem before the problem actually arrives. Second, when the problem arrives, the group may fail to perceive the problem. Then, after they perceive the problem, they may fail even to try to solve the problem. Finally, they may try to solve it but may fail in their attempts to do so. While all this talking about reasons for failure and collapses of society may seem pessimistic, the flip side is optimistic: namely, successful decision-making. Perhaps if we understand the reasons why groups make bad decisions, we can use that knowledge as a checklist to help groups make good decisions.

The first item on my road map is that groups may do disastrous things because they didn't anticipate a problem before it arrived. There may be several reasons for failure to anticipate a problem. One is that they may have had no prior experience of such problems, and so may not have been sensitized to the possibility. For example, consider forest fires in the U.S. West. My wife, my children, and I spend parts of our summers in Montana, and each year when we fly into Montana I look out our plane window as we are coming in to see how many forest fires there are that day. Forest fires are a major problem not only in Montana, but throughout the U.S. Intermontane West in general. Forest fires on that giant scale are unknown in the eastern United States and in Europe. When settlers from the eastern United States and Europe arrived in Montana and a forest fire arose, their reaction was, of course, that they should try to put out the fire. The goal of the U.S. Forest Service for nearly a century was that every forest fire would be put out by 10 a.m. the day after it was reported. That attitude of easterners and Europeans about forest fires was because they had had no previous experience of forest fires in a dry environment where there's a big buildup of fuel, where trees that fall down into the understory don't rot away as in wet Europe and as in the wet eastern United States, but accumulate there in a dry environment. It turns out that frequent small fires burn off the fuel load, and if those frequent small fires are suppressed, then eventually, when a fire is lit, it may burn out of control far beyond our ability to suppress it, resulting in the big disastrous fires in the U.S. Intermontane West. It turns out that the best way to deal with forest fires in the West is to let them burn, and burn out, and then there won't be a buildup of the fuel load resulting in a disaster. But these huge forest fires were something with which eastern Americans and Europeans had no prior experience. The idea that you should let a fire burn and destroy valuable forest was so counterintuitive that it took the U.S. Forest Service decades to realize the problem and to change its strategy and let the fire burn. So here's an example of how a society with no prior experience of a problem may not even recognize the problem—the problem of fuel loads in the understory of a dry forest.

That's not the only reason, though, why a society may fail to anticipate a problem before it actually arises. Another reason is that it may have had prior experience but that prior experience has been forgotten. For example, a nonliterate society is not going to preserve oral memories of something that happened long in the past. The Classic Lowland Maya eventually succumbed to a drought around A.D. 800. There had been previous droughts in the Maya realm, but they could not draw on that prior experience, because although the Maya had some writing, they just preserved the conquests of kings and didn't record droughts. Maya droughts recurred at intervals of 208 years, so when the big drought struck the Maya in A.D. 800, they did not and could not remember the drought of A.D. 592.

In modern literate societies, even though we do have writing, that does not necessarily mean that we can draw on our prior experience. We, too, tend to forget things, and so for example Americans recently have behaved as if they've forgotten about the 1973 Gulf oil crisis. For a year or two after the crisis they avoided gas-guzzling vehicles, then quickly they forgot that knowledge, despite their having writing. And in the 1960s the city of Tucson, Arizona, went through a severe drought, and the citizens swore that they would manage their water better after that, but within a decade or two Tucson was going back to its water-guzzling ways of irrigating golf courses and watering gardens. So there we have a couple of reasons why a society may fail to anticipate a problem before it has arrived.

The remaining reason why a society may fail to anticipate a problem before it develops involves reasoning by false analogy. When we are in an unfamiliar situation, we fall back on reasoning by analogy with old familiar situations. That's a good way to proceed if the old and new situations are truly analogous, but reasoning by analogy can be dangerous if the old and new situations are only superficially similar.

An example of a society that suffered disastrous consequences from reasoning by false analogy was the society of Norwegian Vikings who immigrated to Iceland beginning in the year A.D. 871. Their familiar homeland of Norway had heavy clay soils ground up by glaciers. Those soils were sufficiently heavy that, if the vegetation covering them was cut down, they were too heavy to be blown away. Unfortunately for the Viking colonists of Iceland, Icelandic soils were as light as talcum powder. They arose not through glacial grinding, but through winds carrying light ashes blown out in volcanic eruptions. The Vikings cleared the forests over those soils in order to create pasture for their animals. Unfortunately, the ash that was light enough for the wind to blow in was light enough for the wind to blow out again when the covering vegetation had been removed. Within a few generations of the Vikings' arrival in Iceland, half of Iceland's topsoil had eroded into the ocean. Other examples of reasoning by false analogy abound.

The second step in my road map, after a society has anticipated or failed to anticipate a problem before it arises, involves a society's failing to perceive a problem that has actually arrived. There are at least three reasons for such failures, all of them common in the business world and in academia. First, the origins of some problems are literally imperceptible. For example, the nutrients responsible for soil fertility are invisible to the eye, and only in modern times measurable by means of chemical analysis. In Australia, Mangareva, parts of the U.S. Southwest, and many other locations, most of the nutrients had already been leached out of the soil by rainfall. When people arrived and began growing crops, those crops quickly exhausted the remaining nutrients, so that agriculture rapidly failed. Yet such nutrient-poor soils often bear lush-appearing vegetation; it's just that most of the nutrients in the ecosystem are contained in the vegetation rather than in the soil, so the nutrients are removed when one cuts down the vegetation. There was no way that the first colonists of Australia and Mangareva could perceive the problem of soil nutrient exhaustion.

An even commoner reason for a society's failing to perceive a problem is that the problem may take the form of a slow trend concealed by wide up-and-down fluctuations. The prime example in modern times is global warming. We now realize that temperatures around the world have been slowly rising in recent decades, due in large part to changes in the atmosphere caused by humans. However, it is not the case that the climate each year is inexorably 0.17 degrees warmer than in the previous year. Instead, as we all know, climate fluctuates up and down erratically from year to year: three degrees warmer in one summer than the previous summer, then two degrees warmer the next summer, down four degrees the following summer, down another degree the next summer, then up five degrees, etc. With such wide and unpredictable fluctuations, it takes a long time to discern the upward trend within that noisy signal. That's why it was only a few years ago that the last professional climatologist previously skeptical of the reality of global warming became convinced. President George W. Bush is still not convinced of the reality of global warming, and he thinks that we need more research. The medieval Greenlanders had similar difficulties in recognizing that the climate was gradually becoming colder, and the Maya of the Yucatán had difficulties discerning that the climate was gradually becoming drier.

Politicians use the term "creeping normalcy" to refer to such slow trends concealed within noisy fluctuations. If a situation is getting worse only slowly, it is difficult to recognize that this year is worse than last year, and each successive year is only slightly worse than the year before, so that one's baseline standard for what constitutes "normalcy" shifts only gradually and almost imperceptibly. It may take a few decades of a long sequence of such slight year-to-year changes before someone suddenly realizes that conditions were much better several decades ago, and that what is accepted as normalcy has crept downward.

The remaining frequent reason for failure to perceive a problem after it has arrived is distant managers, a potential problem in any large society. For example, today the largest private landowner and the largest timber company in the state of Montana is based not within the state but in Seattle, Washington. Not being on the scene, company executives may not realize that they have a big weed problem on their forest property.

Jared Diamond

All of us who belong to other groups can think of examples of imperceptibly arising problems, creeping normalcy, and distant managers.

The third step in my road map of failure is perhaps the commonest and most surprising one: a society's failure even to try to solve a problem that it has perceived.

Such failures frequently arise because of what economists term "rational behavior" arising from clashes of interest between people. Some people may reason correctly that they can advance their own interests by behavior that is harmful for other people. Economists term such behavior "rational," even while acknowledging that morally it may be naughty. The perpetrators are often motivated and likely to get away with their rational bad behavior, because, as winners from a bad status quo, they are typically concentrated (few in number) and highly motivated since they receive big, certain, immediate profits, while the losers are diffuse (the losses are spread over large numbers of individuals) and are unmotivated because they receive only small, uncertain, distant profits from undoing the rational bad behavior of the minority.

A typical example of rational bad behavior is "good for me, bad for you and for the rest of society"—to put it bluntly, selfishness. A few individuals may correctly perceive their self-interests to be opposed to the majority's self-interest. For example, until 1971, mining companies in Montana typically just dumped their toxic wastes of copper and arsenic directly into rivers and ponds because the state of Montana had no law requiring mining companies to clean up after abandoning a mine. After 1971, the state of Montana did pass such a law, but mining companies discovered that they could extract the valuable ore and then just declare bankruptcy before going to the expense of cleaning up. The result has been billions of dollars of clean-up costs borne by the citizens of the United States or Montana. The mining companies had correctly perceived that they could advance their interests and save money by making messes and leaving the burden to society.

One particular form of such clashes of interest has received the name "tragedy of commons." That refers to a situation in which many consumers are harvesting a communally owned resource (such as fish in the ocean, or grass in common pastures), and in which there is no effective regulation of how much of the resource each consumer can draw off. Under those circumstances, each consumer can correctly reason, "If I don't catch that fish or graze that grass, some other fisherman or herder will anyway, so it makes no sense for me to be careful about overfishing or over-harvesting." The correct "rational" behavior is to harvest before the next consumer can, even though the end result is depletion or extinction of the resource, and hence harm for society as a whole.

Rational bad behavior involving clashes of interest also arises when the consumer has no long-term stake in preserving the resource. For example, much commercial harvesting of tropical rain forests today is carried out by international logging companies, which lease land in one country, cut down all the rain forest in that country, and then move on to the next country. The international loggers have correctly perceived that, once they have paid for the lease, their interests are best served by clear-cutting the rain forest on their leased land. In that way, loggers have destroyed most of the forests of the Malay Peninsula, then of Borneo, then of the Solomon Islands and Sumatra, now of the Philippines, and coming up soon of New Guinea, the Amazon, and the Congo Basin. In that case, the bad consequences are borne by the next generation, but that next generation cannot vote or complain.

A further situation involving "rational" behavior and conflicts of interest arises when the interests of the decision-making elite in power conflict with the interests of the rest of society. The elite are particularly likely to do things that profit them but hurt everybody else, if the elite are able to insulate themselves from the consequences of their actions. Such clashes are increasingly frequent in the modern U.S., where rich people tend to live within their gated compounds and to drink bottled water. For example, executives of Enron correctly calculated that they could gain huge sums of money for themselves by looting the company coffers and harming the rest of society, and that they were likely to get away with their gamble.

Failure to solve perceived problems because of conflicts of interest between the elite and the rest of society are much less likely in societies where the elite cannot insulate themselves from the consequences of their actions. For example, the modern country with the highest proportion of its citizens belonging to environmental organizations is the Netherlands. I never understood why until I was visiting the Netherlands a few years ago and raised this question to my Dutch colleagues as we were driving through the countryside. My Dutch friends answered, "Just look around you and you will see the reason. The land where we are now is twenty-two feet below sea level. Like much of the area of Holland, it was once a shallow bay of the sea that we Dutch people surrounded by dikes and then drained with pumps to create lowlying land that we call a polder. We have pumps to pump out the water that is continually leaking into our polders through the dikes. If the dikes burst, of course, the people in the polder drown. But it is not the case that the rich Dutch live on top of the dikes, while the poor Dutch are living down in the polders. If the dikes burst, everybody drowns, regardless of whether they are rich or poor. That was what happened in the terrible floods of February 1, 1953, when high tides and storms drove water inland over the polders of Zeeland Province and nearly two thousand Dutch people drowned. After that disaster, we all swore, 'Never again!' and spent billions of dollars building reinforced barriers against the water." In the Netherlands, the decision-makers know that they cannot insulate themselves from their mistakes, and that they have to make compromise decisions that will be good for as many people as possible.

Those examples illustrate situations in which a society fails to solve perceived problems because maintaining the problem is good for some people. In contrast to that so-called rational behavior, there are also failures to attempt to solve perceived problems that economists consider "irrational behavior": that is, behavior that is harmful for everybody. Such irrational behavior often arises when all of us are torn by clashes of values. We may be strongly attached to a bad status quo because it is favored by some deeply held value that we admire. Religious values are especially deeply held and hence frequent causes of disastrous behavior. For example, much of the deforestation of Easter Island had a religious motivation, to obtain logs to transport and erect the giant stone statues that were the basis of Easter Island religious cults. In modern times a reason why Montanans have been so reluctant to solve the obvious problems now accumulating from mining, logging, and ranching in Montana is that these three industries, formerly the pillars of the Montana economy, became bound up with the pioneer spirit and with Montanan self-identity.

Irrational failures to try to solve perceived problems also frequently arise from clashes between short-term and long-term motives of the same individual. Billions of people in the world today are desperately poor and able to think only of food for the next day. Poor fishermen in tropical reef areas use dynamite and cyanide to kill and catch reef fish, in full knowledge that they are destroying their future livelihood, but they feel that they have no choice because of their desperate short-term need to obtain food for their children today. Governments, too, regularly operate with a short-term focus: They feel overwhelmed by imminent disasters, and pay attention only to those problems on the verge of explosion, feeling that they lack time or resources to devote to long-term problems. For example, a friend of mine who is closely connected to the current federal administration in Washington, D.C., told me that, when he visited Washington for the first time after the 2000 national elections, the leaders of our government had what he termed a "ninety-day focus": They talked about only those problems with the potential to cause a disaster within the next ninety days. Economists rationally justify these irrational focuses on short-term profits by "discounting" future profits. That is, they argue that it may be better to harvest a resource today than to leave some of the resource for harvesting tomorrow, because the profits from today's harvest could be invested, and the accumulated interest between now and a harvest of exactly that same quantity of resource in the future would make today's harvest more valuable than the future harvest.

The last reason that I shall mention for irrational failure to try to solve a perceived problem is psychological denial. This is a technical term with a precisely defined meaning in individual psychology, and it has been taken into pop culture. If something that you perceive arouses an unbearably painful emotion, you may subconsciously suppress or deny your perception in order to avoid the unbearable pain, even though the practical results of ignoring your perception may prove ultimately disastrous. The emotions most often responsible are terror, anxiety, and sadness. Typical examples include refusing to think about the likelihood that your husband, wife, child, or best friend may be dying, because the thought is so painfully sad, or else blocking out a terrifying experience. For example, consider a narrow, deep river valley below a high dam, such that if the dam burst, the resulting flood of water would drown people for a long distance downstream. When attitude pollsters ask people downstream of the dam how concerned they are about the dam's bursting, it's not surprising that fear of a dam burst is lowest far downstream, and increases among residents increasingly close to the dam. Surprisingly, though, when one gets within a few miles of the dam, fear of the dam's breaking is highest; as you then get closer to the dam the concern falls off to zero! That is, the people living immediately under the dam, who are certain to be drowned in a dam burst, profess unconcern. That is because of psychological denial: The only way of preserving one's sanity while living immediately under the high dam is to deny the finite possibility that it could burst.

Psychological denial is a phenomenon well established in individual psychology. It seems likely to apply to group psychology as well. For example, there is much evidence that, during World War Two, Jews and other groups at risk of the developing Holocaust denied the accumulating evidence that it was happening and that they were at risk, because the thought was unbearably horrible. Psychological denial may also explain why some collapsing societies fail to face up to the obvious causes of their collapse.

Finally, the last of the four items in my road map is the failure to succeed in solving a problem that one does try to solve. There are obvious possible explanations for this outcome. The problem may just be too difficult, and beyond our present capacities to solve. For example, the state of Montana loses hundreds of millions of dollars per year in attempting to combat introduced weed species, such as spotted knapweed and leafy spurge. That is not because Montanans don't perceive these weeds or don't try to eliminate them, but simply because the weeds are too difficult to eliminate at present. Leafy spurge has roots twenty feet deep, too long to pull up by hand, and specific weed-control chemicals cost up to $800 per gallon.

Often, too, we fail to solve a problem because our efforts are too little, begun too late. For example, Australia has suffered tens of billions of dollars of agricultural losses, as well as the extinction or endangerment of most of its native small mammal species, because of the introduction of European rabbits and foxes for which there was no close native counterpart in the Australian environment. Foxes as predators prey on lambs and chickens and kill native small marsupials and rodents. Foxes have been widespread over the Australian mainland for over a century, but until recently they were absent from the Australian island state of Tasmania, because they could not swim across the wide, rough seas between the mainland and Tasmania. Unfortunately, two or three years ago some individuals surreptitiously and illegally released thirty-two foxes on the Tasmanian mainland, either for their fox-hunting pleasure or to spite environmentalists. Those foxes represent a big threat to Tasmanian lamb and chicken farmers, as well as to Tasmanian wildlife. When Tasmanian environmentalists became aware of this fox problem around March 2002, they begged the government to exterminate the foxes quickly while it was still possible. The fox breeding season was expected to begin around July. It would be far more difficult to eradicate 128 foxes than thirty-two foxes, once those thirty-two foxes had produced litters and those litters had dispersed. Unfortunately, the Tasmanian government debated and delayed, and it was not until around June 2002 that the government finally decided to commit a million dollars to eliminating foxes. By that time, there was considerable risk that the commitment of money was too little and too late, and that the Tasmanian government would find itself faced with a far more expensive and less soluble problem. I have not heard yet what happened to that fox eradication effort

Thus, human societies and smaller groups may make disastrous decisions for a whole sequence of reasons: failure to anticipate a problem, failure to perceive it once it has arisen, failure to attempt to solve it after it has been perceived, and failure to succeed in attempts to solve it. All this may sound pessimistic, as if failure is the rule in human decision-making. In fact, of course that is not the case, in the environmental arena as well as in business, academia, and other groups. Many human societies have anticipated, perceived, tried to solve, or succeeded in solving their environmental problems. For example, the Inca Empire, New Guinea Highlanders, 18th-century Japan, 19th-century Germany, and the paramount chiefdom of Tonga all recognized the risks that they faced from deforestation, and all adopted successful reforestation or forest management policies.

Thus, my reason for discussing failures of human decision-making is not my desire to depress you. Instead, I hope that, by recognizing the signposts of failed decision-making, we may become more consciously aware of how others have failed, and of what we need to do in order to get it right.

Jared Diamond

3.

Art and Human Reality

Denis Dutton (1944–2010)

Philosopher; founder, Arts & Letters Daily; *author,*
The Art Instinct: Beauty, Pleasure, and Human Evolution

INTRODUCTION BY STEVEN PINKER

*Johnstone Family Professor, Department of Psychology, Harvard
University; author,* The Language Instinct, The Blank Slate,
and The Stuff of Thought

*Denis Dutton is a visionary. He was among the first (together with
our own John Brockman) to realize that a website could be a forum for
cutting-edge ideas, not just a way to sell things or entertain the bored.
Today,* Arts & Letters Daily *is the website that I try the hardest not to
visit, because it is more addictive than crack cocaine. He started one of the
first print-on-demand services for out-of-print scholarly books. He saw
that philosophy and literature had much to say to each other, and started
a deep and lively scholarly journal to move that dialogue along. He saw
that pompous and empty prose in the humanities had become an impedi-
ment to thinking, and initiated the Bad Writing Contest to expose it.*

*And now he is changing the direction of aesthetics. Many people be-
lieve that this consilience between the arts, humanities, and sciences represents the future of the humanities, revitalizing them with a progressive
research agenda after the disillusionments of postmodernism. Dutton has
written the first draft of this agenda. He has defended a universal definition of art—something that many theorists assumed was simply impos-*

sible. And he has advanced a theory that aesthetics have a universal basis in human psychology, ultimately to be illuminated by the processes of evolution. His ideas in this area are not meant to be the last word, but they lay out testable hypotheses, and point to many fields that can be brought to bear on our understanding of art.

I see this as part of a larger movement of consilience, in which, to take a few examples, ideas from auditory cognition will provide insight into music, phonology will help illuminate poetics, semantics and pragmatics will advance our understanding of fiction, and moral psychology will be brought to bear on jurisprudence and philosophy. And in his various roles, Denis Dutton will be there when it happens.

What we regard as the modern human personality evolved during the Pleistocene, between 1.6 million and 10,000 years ago. If you encountered one of your direct ancestors from the beginning of the Pleistocene moseying down the street today, you would probably call the SPCA and ask for a crew with tranquilizer darts and nets to cart the beast off to the zoo. If you saw somebody from the end of the Pleistocene, 10,000 years ago, you'd call the Immigration and Naturalization Service—by that time our ancestors wouldn't have appeared much different from any of us today. It is that crucial period, those 80,000 generations of the Pleistocene before the modern period, which is the key to understanding the evolution of human psychology. The features of life that make us most human—language, religion, charm, seduction, the seeking of social status, and the arts—came to be in this period, no doubt especially in the last 100,000 years.

The human personality—including those aspects of it that are imaginative, expressive, and creative—cries out for a Darwinian explanation. If we're going to treat aspects of the personality, including the aesthetic expression, as adaptations, we've got to do it in terms of three factors.

The first is pleasure: The arts give us direct pleasure. A British study a few years ago showed that 6 percent of all waking life of the average British adult is spent enjoying fictions, in movies, in plays, and on television. And that didn't even include fictional books—bodice-rippers, airport novels, high literature, and so forth. That kind of devotion of time and its pleasure-payoff demands some kind of explanation.

As a second comes universality. What we've had over the past forty years is an ideology in academic life that regards the arts as socially constructed and therefore unique to local cultures. I call it an ideology because it is not argued for, it is just presupposed in most aesthetic discourse. Allied with this position is the idea that we can seldom or perhaps never really understand the arts of other cultures; other cultures likewise can't understand our arts. Everybody's living in his or her own socially constructed, hermetically sealed, special cultural world.

But of course, a moment's thought reveals that this can't possible be true. We know people in Brazil love Japanese prints, that Italian opera is enjoyed in China. Both Beethoven and Hollywood movies have swept the world. Think of it—the Vienna Conservatory has been saved by a combination of Japanese, Korean, and Chinese pianists. The universality of the arts is a fact, again, a fact that requires explanation. We simply can't keep going on forever making this false claim that the arts are unique to cultures.

And third, we have to consider the spontaneity of the arts—the way they spontaneously arise, beginning in childhood experience, across the globe. Think of the ways in which children, by the time they're three years old, can engage in make-believe and keep imaginary worlds separate from one another. A small child is playing with its teddy bears at a tea party. If you knock over a cup and spill the pretend tea in it, the child will not be in the least confused as to which of the three empty cups to refill. In fact, if you refill the wrong empty cup, and insist it was the one that spilled, the child may well break out in tears. The child then goes from the tea party over to the television and watches a Bugs Bunny cartoon or *Sesame Street*. From there, it's on to reading a book, entering into its make-believe world, and then to having dinner with Mommy and Daddy. Even a three-year-old can keep all of these real and fictional worlds coherently separate from one another. Such spontaneous intellectual sophistication—try to imagine teaching it from scratch to a three-year-old—is a mark of an evolved adaptation.

Pleasure, universality, spontaneous development. We see them in the cross-cultural realities of music and the universality of storytelling, as well as in things like food tastes, erotic interests, pet-keeping, sports interests, our fascination with puzzle-solving, gossip—the list is indefinitely long. Charles Darwin has a lot more to say about how we evolved as inventive and expressive social animals with our remarkable personalities than he has been given credit for. These aspects of evolution have deep implications for the origins and evolution of the arts.

You ask why I have such a long-standing interest in the genesis of artistic experience. I don't really know. I grew up in Southern California; my parents had met at Paramount Pictures, where they worked in the 1930s. They later founded bookstores, the Dutton Books of Southern California. I think that among my earliest memories must be sitting on the living room floor playing over and over again a recording of Beethoven's Seventh Symphony. To my child's mind, this music was magical, its pleasure intense.

I took violin and piano lessons as a child, but was never very good with anything I could not memorize. I seem to have some mildly dyslexic inability to read music fluently, though my musical memory is fairly prodigious—I do know that standard run of Western classical music inside and out.

I entered the University of California–Santa Barbara, originally as a chemistry major, but soon changed to philosophy and was fascinated by aesthetics. As an undergraduate, I was taught—and more or less accepted—elements in Wittgenstein and anthropology that proclaimed the uniqueness and incommensurability of cultures and art forms.

It's not as though this was ever backed up by serious arguments. It was supported by anecdotes. My generation was taught that the Eskimos had five hundred words for snow. It's an urban legend; it's simply not true. But if you believed it, then you could believe that the Eskimo lives in a special intellectual world of which we're not a part.

Consider the story, equally fabulous, about the African who, for the first time shown a photograph of a person, didn't know how to read it as a photograph, couldn't see it as a representation of a person. Fancy that: The confused African couldn't see any natural resemblance between a photograph and a live person. My experience in New Guinea would indicate that's just ridiculous. I can imagine that the African might have been a bit confused when for the first time he saw a truck come into his village, a white man getting out of it and shoving a piece of paper in front of his face. But to turn such an incident into a failure to understand a naturalistic representation—that's just loopy social constructionist ideology; it's not serious research on what were then called "primitive" cultures.

Another one of my favorite myths is the story of Ravi Shankar in San Francisco giving a concert. He comes out on stage and tunes the sitar. Now the sitar is a very complicated instrument to tune, and he works on it for about ten minutes. When he's finished, he nods to the audience and everybody applauds thinking that was actually the first piece of music on the program. Ipso facto, people cannot really understand foreign cultures.

After I got out of college, I joined the Peace Corps and went to South India. I worked in a village north of Hyderabad. It was a Dravidian-language-speaking culture with the caste system of India, in many ways ancient and very foreign—Southern California it was not. On the other hand, if you looked at the foibles and passions and absurdities and ambitions and plans that people have for their lives, Indian culture was completely intelligible.

Indians are not another species of animal. They're human beings, and we can understand them. And I found out we can understand their music, because I started playing the sitar in India, studying with Pandurang Parate, a student of Ravi Shankar himself. I still play the sitar. In fact, I can get free meals in Indian restaurants in the town where I live by twanging on the sitar for a while for entertainment. I've played it on and off for forty years.

And by the way, I found out what was behind that story about Ravi Shankar in San Francisco. It's another urban legend concocted to support the thesis that cultures can't understand each other. No one who has watched the sitar being tuned could possibly think that fiddling with the pegs and the strings is a piece of music. No San Francisco audience, no matter how stoned, could mistake that for a performance: The applause was just relief that the tedious tuning was finished.

But the story got incorporated into the 1960s zeitgeist. It's time to be done with these fables, after forty or fifty years, and ask ourselves why the arts are universal. The notion that art is purely socially constructed, indeed, the human personality is socially constructed, has to make way for something more complex.

After grad school at NYU and UCSB, I taught philosophy at the University of Michigan–Dearborn, and later moved to New Zealand, where for some years I taught philosophy of art in my university's school of fine arts. I taught courses across the board in philosophy—the history of philosophy and various subdivisions of philosophy, but the nagging aesthetics questions persisted. My colleagues all seemed to agree that culture was the only way to explain art, but this position seemed unsatisfactory.

In the late 1980s, I developed a passionate interest in oceanic art and the carvings of New Guinea. One day, my wife suggested, "Well, we're close enough. Why don't you simply go up to New Guinea and find out what their aesthetic standards are?" By that time, I was well acquainted with what European connoisseurship would call the "greatest" works of New Guinea art. But would the European valuations accord with local New Guinean valuations? Australian friends, old New Guinea hands, helped me to find a village, Yentchenmangua, on the Sepik River, where carving traditions were still alive. (This project had an unintended by-product: Somewhere out there in a museum or gallery there's an authentic New Guinea carving carved by me. I'd left one of my practice carvings in the village and only found out later that it had been painted and sold off.) This experience taught me something crucially important: that New Guinea standards for greatness and for excellence are, as far as I could determine, the same as those of knowledgeable European curators, connoisseurs, and collectors.

I'm not saying that the New Guineans would make judgments that would coincide with every naive tourist—a newcomer to the art—who gets off the boat. Tourists in my experience make very bad choices in buying New Guinea art. But the people who really know the good work in museums, who are very deeply familiar with New Guinea art but who have never set foot in New Guinea, oddly have the same taste patterns as New Guinea carvers themselves. And this shows that with the art form, knowledge and familiarity with the whole field determines a convergence of taste. And that, again, has to be explained.

You could try to explain it by saying that God has imprinted us with something. Jung thought he had ways of approaching this. Joseph Campbell was interested in these issues. But the person who really has the answers is Charles Darwin. In his first books, which are amazingly detailed, he couldn't go into all of these specific aesthetic issues, but he set out the blueprint for us. And we can apply Darwinian ideas and come to some initial rough account. I hope that over the years my arguments about the genesis of artistic taste will be refined.

And I have to stress that I am far from claiming that I have all the answers about the evolutionary origins of aesthetic taste. Darwinian aesthetics is not some kind of ironclad doctrine that is supposed to replace a heavy post-structuralism with something just as oppressive. What surprises me about the resistance to the application of Darwin to psychology is the vociferous way in which people want to dismiss it, not even to consider it. Is this a holdover from Marxism or religious doctrines? I don't know. Stephen Jay Gould was one of those people who had the idea that evolution was allowed to explain everything about me, my fingernails, my pancreas, the way my body is designed—except that it could have nothing to say about anything above the neck. About human psychology, nothing could be explained in evolutionary terms: We just somehow developed a big brain with its spandrels and all, and that's it.

Denis Dutton

This position is unsupportable. We know there are built-in spontaneous features of the human personality, conspicuously present, for instance, in the evolutionary development of speech. But other aspects of the personality as well, ones having to do with the arts, are also universal, appearing in childhood with little or no prompting, or simply arising "naturally," so it seems to us, as features of social interactions.

I cannot understand why there still is so much resistance among academics to such ideas. If you want to be a one-dimensional determinist, go ahead and make it all "culture." My side of the argument isn't trying to make it all "nature," make it all genetics. Human life is lived in a middle position between our genetic determinants on the one hand and culture on the other. It's out of that that human freedom emerges. And artistic works—the plays of Shakespeare, the novels of Jane Austen, the works of Wagner and Beethoven, Rembrandt and Hokusai—are among the freest, most human acts ever accomplished. These creations are the ultimate expressions of freedom.

It makes no more sense to claim that our artistic and expressive lives are determined only by culture than it does to say that we are determined only by genes. Human beings are a product of both. Why can't we get over our post-Marxist nostalgia for economic or cultural determinism and accept human reality as it actually is? The truth of the human situation is that we are biologically determined organisms that live in a culture. That we are cultural creatures is part of what is determined by our genes.

It's a great question, What is art? But it's been answered in the wrong way by philosophers for the past forty years. The fundamental mistake has been to imagine that if we could explain why Duchamp's great work *Fountain* is a work of art, then we'd know what traditional works of art are. I say no to this procedure. Instead of asking how is it that Duchamp's readymades are works of art, I say, Let's ask what is it that makes the *Pastoral* Symphony a work of art. Why is *A Midsummer Night's Dream* a work of art? Why is *Pride and Prejudice* a work of art? Let's look first at the undisputed paradigm cases and find out what they all have in common—and not only in the Western tradition but also in the great Eastern traditions of China and Japan. Look at Hokusai, consider New Guinea carving, and look at African carving. Better to understand them first, and then analyze modernist experimentation and provocations, such as Duchamp's brilliant work. I do regard Duchamp as an incandescent genius. But our respect for him must include a recognition of the fact that he was in some of his works experimenting in ways intended to outrage and provoke people by implicitly asking what the limits of art are.

To put the point analogically: If you're teaching ethics in a philosophy class and you want to get to understand what murder is, you don't begin by asking whether capital punishment or abortion or assisted suicide is murder. What you do is start with the clear cases and then move out later to ask, Is capital punishment murder? We ought first to make sense of the clear cases.

An obsession with marginal cases has actually degraded the discussion in aesthetic theory of what the arts are. I must say it's made for a lot of fun in philosophy of art classes. Duchamp's gestures are sure to get students interested. It's the same with questions like, What is wrong with a forgery? Or, Is there an intentional fallacy in interpreting literature? These issues generate intriguing conundrums. But after we've had our fun, we must also get back to central questions of what it is that makes the *Iliad* or *Guernica* art. Then we can better deal with Duchamp.

————

Denis Dutton

Modernism has long had a project—to oversimplify—directed against the excesses, pomposity, and absurdities of the 19th-century art that preceded it. Think of those huge, gaudy, sentimental paintings produced by the Victorians. You'll find many in the basements of art galleries and museums in New Zealand: gigantic canvases of biblical themes—*The Flight into Egypt*, perhaps. Many of these paintings cannot be regarded today as anything but big dark monstrosities, and white elephants so far as storage space goes. No one wants to look at them—but no one knows what to do with them, either.

We're in the same situation right now with the late 20th century and the beginning of the 21st century. Our museums are burdened with gigantic mega-canvases. Will anyone be interested in seeing them in a hundred years? Will anyone actually care about a shark in formaldehyde in a hundred years? (That's a particularly tough one: Even in formaldehyde, that shark will likely have disintegrated in a hundred year's time. Or is that in fact part of the whole work of art?) This is an interesting issue. I'm not sure I want to put it in permanent storage, or the huge canvases produced in the 1970s, where size alone was supposed to prove it was great art. Well, it didn't then, and it still doesn't now.

Many times in its history, including ours, art has experienced periods of folly. It's fun to watch, of course, but as a Darwinian I'm also interested in the features of works of art that are going to make them still be looked at and listened to and read five hundred years from now. That for me is the question. By the way, I think that Warhol stands a chance, as does Jackson Pollack. On the other hand, I'm not so sure about Schoenberg, particularly his atonal music.

Anton Webern once suggested that someday we will have advanced to the point where the postman will in his sophistication do his rounds whistling an atonal non-tune. A lovely hope for modernism, but the idea is completely implausible. What is it about a melody that a Schoenberg tone row doesn't quite qualify in the minds of most people? That's a question about basic human musical psychology. And, of course, it's the reception of twelve-tone music that is usually presented as though it's a question about culture—or resistance to change. I don't think it's about culture. Alone.

One of the earliest influences on my thinking on this was Ellen Dissanayake, who has written three major books and a lot of articles. She wrote a book entitled *What Is Art For?* and then *Homo Aestheticus*. Her most recent book is *Art and Intimacy*. Her view of the arts was a revelation. She wasn't trying to disparage the arts, reduce them to a brute drive, or make them any less than the grand things they are. She did want to connect them with an evolved human nature in a way that makes a lot of sense. One of the great ironies of the academic world is that this woman who has made such a contribution with her books and articles has never been able to land an academic job. She's a medical stenographer in Seattle. After working all day, she goes home to write at night or on the weekends pioneering books on the subject of evolutionary aesthetics. I regard her as one of the most remarkable intellectual figures of our time.

Of course, John Tooby and Leda Cosmides are extremely important in their groundbreaking work in evolutionary psychology. Steven Pinker is so imaginative and informed: He has been a great inspiration. Joseph Carroll has done exquisitely sophisticated research in literary Darwinism, sometimes alongside his younger colleague, Jonathan Gottschall. Brian Boyd, my colleague in New Zealand famed for his Nabokov biography, is also heavily involved in the evolutionary psychology of literature.

These people have meant a lot to me and have helped me to overcome, if I may say so, my own Wittgensteinian enculturation in which forms of life are considered incommensurable between cultures. It's not just Foucault and Derrida: Wittgenstein also has a lot to answer for. There's a deep anti-naturalism in his work, but a consistent ambiguity that makes it difficult to identify. Consider Wittgenstein's gnomic, seemingly profound claim: "If a lion could speak, we could not understand him." Oh, yeah? That's a deeply mischievous idea, and Wittgenstein would have profited from getting to know an animal ethologist or two. If a lion could speak, the ethologists would be pretty clear about what he'd be talking about: annoying other lions, members of the opposite lion sex, tasty zebras, and so on. People who live with animals can understand them, sometimes rather remarkably.

On the other hand, used in the wrong ways, animal ethology can itself be misleading. In evolutionary aesthetics, animals have to be used to explain evolutionary principles, natural selection and sexual selection in the human situation. Take, for instance, chimpanzee art. We became human in the Pleistocene, having forked off from the chimps fully 5 million years before that—which means we are still very distant indeed from our closest surviving primate relatives. These days, people in zoos and primate research centers enjoy taking out big sheets of butcher paper and letting the chimps go at them with brushes and paint. The chimps have a grand old time, scribbling about or making a typical upward fan figure. They are essentially taking joy in the sheer disruption of the white background with a solid color. It's not unlike the pleasure many of us have gotten with finger painting, or early painting in school: We can get pleasure simply in the contrasts that we create.

Is this "chimpanzee art"? People who make such claims are

usually not aware of other aspects of the chimps' behavior. First, the typical upward fan shape actually is not a picture, an image of a fan, because a chimp can't turn it on its side or render it upside down. It's not a representation so much as part of a motor sequence in the chimp's arms and hands. Second, if the trainer does not take the piece of paper away from the chimp, the result will inevitably be a brownish blob, because a chimp has no idea of when to stop. There is no objective, or sense of a plan or end point in creating the work. It's only a work of art for us because the trainer took it away from the chimp before it became a blob. Finally, and for me most tellingly, when they're finished—or the paper's been taken away—the chimps never again go back to look at the work.

It seems to me that anyone who says, "Yes, chimpanzees have art," is making a mistake. Chimpanzees like to disrupt white paper with big colored blobs. As human beings, we can understand that, but that does not make their creations works of art. There is no cultural tradition within which chimps are working. There's no criticism—art talk or evaluation of any kind—with the chimps. There's no style in the sense that it's a learned way of doing it, though there are uniformities in the output for muscular reasons. To call this art or proto-art underestimates and misunderstands what human art is.

Animals have much to teach us, but from a Darwinian perspective, human beings really are something else.

Denis Dutton

4.

A Big Theory of Culture

Brian Eno

Artist; composer; recording producer: U2, Coldplay,
Talking Heads, Paul Simon; recording artist; author,
A Year with Swollen Appendices

INTRODUCTION BY STEWART BRAND

Cofounder and cochairman, The Long Now Foundation; author,
Whole Earth Discipline

Here's what I greatly appreciate about Brian Eno, apart from the plea-
sure I take from his friendship and from the pure delight of his music
and art . . .

Like all significant artists, Brian works from a deep and complex and
evolving frame of reference. Unlike most artists, and like most scientists,
he talks about that frame of reference. He's not worried that your experi-
ence of his art might be sullied by your understanding something about
what he's up to—rather the opposite: He would like to include you in the
process.

This is risky, but valuable. It's risky because once viewers or listeners
know what the artist is attempting, they have criteria for judging when
he has failed.

Brian's approach is valuable because it is so inviting. The informed
viewer or listener is invited to think like an artist and therefore in a sense
to become an artist. This is good for art and good for civilization.

I think that's what makes Brian's book A Year with Swollen Ap-

pendices *so appealing. Brian is famous, and that makes us interested, and he's charming in print as well as in person, so we engage with him comfortably. But what gets us about the book is how revealing it is. We see what a good artist does with his mind all day. It's inspiring.*

There's a further benefit to telling all, this time to the artist. By not keeping his frame of reference secret, Brian is freed from binding allegiance to whatever he was thinking when he first became successful. You don't cling to secrets you've told. You move on, and your work keeps being surprising as a result. Maybe this approach works best with artists who are easily bored. Brian is, after all, the author/composer/performer of the tune (now a well-known meme) "Been There, Done That."

EDGE: Let's talk about your theory of culture.

BRIAN ENO: I guess the question I've always been really interested in, the one that underlies all the others, is alluded to somewhat in my book, and I've written about it more since, which is to try to find a big theory about culture: why people do culture, what it does for us, what we actually call culture, which things do we include in that category, and which things do we leave out. I have two intentions in thinking about this. One is that I want to find a single language within which one can talk about fashion, cake decoration, Cézanne, abstract paintings, architecture—within which one can discuss anything one might call nonfunctional, stylistic behavior—which is what humans actually spend more and more of their time doing. The better off humans are, the more time they spend engaged in issues of style, essentially—making choices between one look of things and another look of things. The first question is to say, Is there one language within which we can talk about all of those things? There doesn't have to be a separate language for fine art, so-called, separate from anything else we talk about. There should be one language that fits these things together.

The second question is to try to say, Is there a way of understanding why humans continuously and constantly and without exception engage in cultural activity? We don't know of human groups that don't produce something that we would call art. It seems to be something that we are biologically inclined to do. If we are, then what is the nature of that drive? What is it doing for us? When people say, well, surely this has been written about, what I say is, actually, it hasn't, really. The number of books on this subject is vanishingly small. They occupy a shelf about eighteen inches long. What has been done is a huge sort of taxonomy of cultural artifacts; people sort of listing things and saying, that looks a bit like that, and these seem to belong together, and so on and so on. But I always say that this is a little bit like natural history before Darwin came along. Before Darwin, there were lots of observations, there were people noticing that all of these things existed, making careful notes about them, talking about them, saying that this related to this, this was higher than that or lower than that, and making all the sorts of judgments and observations that people now make about cultural behavior. When Darwin came along, what he said was very simple, very easy for anyone to understand and extremely profound, because it gave one language: the language of survival and the drive to survival and selection and so on. He gave one language in which one could frame all of the things called living organisms. By doing that, it made that subject not just a way of collecting heaps of material, but of actually making theories about that material. In a way, he brought to an end the sort of gathering stage of natural history, the stage where the job of a natural historian was just to go out and make observations, and he brought into being the next phase, which was the task of somehow relating things together and making extrapolations and predictions, and saying, if this happens, we might expect that would happen. That's the job of science.

EDGE: But you're an artist. Why are we talking about Darwin?

ENO: Most of the questions I'm interested in about art and culture really are based on trying to look at them with some kind of big theory of that kind, which is not oblique, not mysterious, is quite easily graspable, and would allow a real discussion about culture. It's partly because I think most art writing is absolutely appallingly bad.

My first mother-in-law, that's to say the mother of my first wife, was a very interesting woman who lived in Cambridge and had a salon at which quite a lot of very good scientists would appear, Francis Crick, John Kendrew, and Hermann Bondi, among others. Her name was Joan Harvey and she ran a thing called the Cambridge Humanists. I met her daughter, and was taken home, and got along very well with Joan. I was seventeen at the time. One day Joan said to me, It's all very well what you do, but I just don't understand why someone with a brain as good as yours wants to waste it being an artist. This question cut me to the quick in a way. I came from a working-class background where nobody particularly cared what you did. It was the first time that anyone had ever cared. Then I fell in with a lot of arty people, who of course assumed that being an artist was a wonderful thing, and never bothered to ask the question about why—about what the point of it might be, or what it actually did for anybody. Joan asked that question, and I never stopped thinking about it. That was the beginning of an interesting double life, because part of my life of course is being an artist, but the other part, and just as interesting to me, is wondering what it is I'm doing, or what everybody else is doing—asking what it's for.

EDGE: How do you think the arts and the sciences differ?

ENO: If you asked twenty scientists what they thought they were doing, or what they thought the point of science was, I would think that most of them would come up with an answer something like, We want to understand the world, we want to see how the world works. If you asked twenty artists the same question—what are you doing it for, what does art do for us—I guarantee you'd get about fifteen different answers, and the other five would tell you to mind your own business. There is no consensus whatsoever about what art is there for, although some people will say, Well, it's to make life more beautiful.

Here I am, an artist—who reads mostly science books—like most other artists. I know very few artists who read books about art. Why, I ask myself, is there not a conversation of that quality in the arts? Many artists normally are talking about science, they're not talking about art—there is not a developed language for having a conversation about the arts.

I'm gradually arriving at some sort of a theory of culture that is getting a few adherents now. I've been talking about it awhile, and I've slimmed it down enough that it is communicable in less than two days.

The first assumption is that all human groups engage in something that we would call artistic behavior—if they are at all capable of it, that is if they are beyond the most basic problems of survival, and even when they aren't, they will engage in decorative, ornamental, and often very complex stylistic behavior. This takes a big chunk of their resources. It takes a lot of energy. So the first question is, Why would that be the case? If it is the case, one would assume that it's doing something more than just mildly entertaining—it's doing something important for us.

The second assumption is this thing I mentioned earlier about assuming that culture is in some sense a unified field, in the same sense that life is a unified field. So that one wants to come up with a language, just as biologists want to come up with a language within which you can discuss whales and amoebas without having to invent a whole new set of terms for each of them. You want to have some structure underneath that would say, yes, we can locate those things within the same pantheon of possibilities.

EDGE: So is this an artistic analogue to a unified field theory?

ENO: I want to find a way of talking about culture, so therefore if I talk about it, I have to be able to include everything from what's considered the most ephemeral, menial, and unimportant version of culture—haircuts, shoe designs—to what are considered the most hallowed and eternal examples of it. Now when I try to think about what it does for us, I try to think what happens to you in certain specific situations. For example, let's take this pair of designer sunglasses that happen to be on the table in front of me. They're very styled. They don't have to be like that. Glasses don't have to be funny, oval, weird-shaped looking glasses, space-age-type glasses. As I put those glasses on, I'm not only keeping the sun out of my eyes. I'm also engaging in some kind of game with myself and the rest of the world. What I'm doing is I'm entering into some kind of simulator. I'm saying, What would it be like to be the kind of person who wears these kinds of glasses? What I mean by that is, I'm not actually abandoning who I am and becoming somebody else; I'm for a while entering into a game where I suddenly become this person who's a different person from the person you've just been talking to.

With all fashion, what we do is play at being somebody else. We play at inhabiting another kind of world. If I decide to cut my hair short and dress like a tank commander, I play with the resonances of kitsch, militaria, dominance, and surrender and control,

Brian Eno

and strength and weakness and all those sorts of things—I'm role-playing, effectively, when I'm making fashion choices.

If I go to a cinema and I look at a film, what I do is take part in another kind of role-playing. I first of all watch a world being constructed, and if the film is any good, I understand what the conditions and rules of that world are, and then I watch a few people who represent certain sets and bundles of characteristics, and I see what they do and how they relate to that world. Essentially what I'm watching is a kind of experiment that's been set up. I'm watching what it would be like if the world was like this, and what it would be like if this kind of person met that kind of person in that kind of context.

EDGE: Is this something one does consciously?

ENO: This kind of playing with other worlds, this ability to move from the world in my head to the possible world in your head, and all the other millions of possible worlds that we can imagine, is something that humans do with such fluency, and such ease, that we don't notice ourselves doing it. We only notice how powerful that process is when we meet people who can't do it—severely autistic children, for instance, who are incapable of switching worlds, and who in many senses can appear completely intelligent but are incapable of seeing that there is any world other than the one that they perceive at this moment. This makes them incapable of two very important things: They can't cooperate easily, because to cooperate you have to understand not only your world, but the world of the person with whom you're cooperating, because you're trying to make a new common world, so you have to see where the other two worlds are. And they can't deceive. Severely autistic children also are incapable of deception, because they could not understand how they could create a situation in which you could see a different world from the one that they believe exists.

To a very large degree, cooperation and deception are the two things that distinguish human beings from the other animals. We have noticed now that some of the higher primates have the rudiments of cooperation and deception, but compared to ours, they really are very rudimentary. My argument is that what the constant engagement in culture does for us is that it enables us to continually rehearse this ability we have—the use of this big part of our brain that is involved in postulating, imagining, exploring, and extrapolating other worlds, either individually or cooperatively.

This is the point at which there is a deep connection between art and science: Each is a highly organized form of pretending, of saying, Let's see what would happen if the world were like this.

EDGE: Let's move on to your ideas about metaphors.

ENO: "The other worlds theory," you might say, is one part of my idea. The other part is what I call "the metaphors theory." Humans actually codify most of their knowledge not in terms of mathematical tables, sets of statistics and scientific laws, but in terms of metaphors. Most of the things we normally have to deal with understanding are complex, fuzzy, messy, changing, and in fact poorly delineated. We don't actually know where the boundaries of them are, let alone whether we are able to make clear questions about them. We spend a lot of our time as ordinary humans navigating through complicated situations with one another that require constant negotiation and constant new attempts to understand.

Science is, of course, one extreme version of this process. Science works by trying to say, Okay, I can separate off this piece of the world from the rest. Effectively, we can say, I've separated that off, and then I can make some theories and predictions about it. Science therefore enables us to come up with a structure upon which we can build useful metaphors. This is why artists are interested in science—it's because science keeps coming up with big ideas, like chaos, like complexity, that we then think, ah, yes, perhaps that's how a lot of things work. Then we have a new metaphor. We don't have to fully understand the science that made that metaphor.

A lot of those kinds of metaphors derive from science, but a lot of them derive from literature, poetry, music. We live in a big construction of metaphors —nearly all of our knowledge is rather fuzzy in that sense. One of the things that artists do is invent metaphors, break up metaphors, challenge them, pull them apart, put them together in new orders and so on. One of the things art does also is to remind you constantly of this process that you're most of the time engaged in—the process of metaphor-making. I am interested in the work of George Lakoff. I thought that *Metaphors We Live By* was a very interesting book, because it pulls you away from the old model of the mind having two departments, the rational department and the kind of intuitive department. It says, no, it's not quite like that. It says there's a continuum, that there are places where we can be strictly rational, such as when I'm doing my accounts with my calculator, when I'm making precise estimates of how I'll make something and what it'll be like. I can use all of the purely rational tools for that. But then there's a whole continuum, which is actually unbroken as far as I can see, where at the one end I can be entirely rational, then I can be pretty logical but I have to make a few guesses, and right down to another end here where it's pure hunch. It's absolutely pure hunch.

EDGE: How does it all come together? Or does it come together at all?

ENO: Mostly we're given the impression that there are just these two separate ways of doing things. However, I believe that one is constantly navigating along that whole spectrum. And that process of navigation is a process of donning different kinds of metaphor, accepting the usefulness of different kinds of metaphor. Once again, this hasn't been really worked on by art writers.

Any of the interesting work on this has been done from the position of science, and has therefore tended to want to address that end of the spectrum of things. If I drew that spectrum of the highly rational to the highly intuitive, what I would have to say is that we don't spend much of our time at either of those extremes. We spend most of our time negotiating somewhere along the middle.

You have art writers who constantly celebrate the "intuition" extreme and think that this is the sort of apex of human existence, and you have scientists who by default, almost, dignify the other one. That's where they live, or that's where they'd like to live. They want to be able to make the kind of statements that push that boundary. What I would like to see is a conversation that admits that we spend most of our time somewhere in the middle, and we ought to find a way of thinking about it.

I suppose at the root of all this is the feeling that possibly the only way that humans can remain cooperative is by those of us who are artists or who are interested in the arts realizing that we have some kind of a job to do. It's no good anymore, as far as I'm concerned, for artists to just take the bohemian attitude of, oh, it just comes out of me, and I don't know what I'm doing, etc. I just can't stand that; I don't want this romantic attitude that says artists shouldn't be part of this planet. This is a real job, and it has to do something.

EDGE: How do you do this job?

ENO: I wrote to Richard Dawkins, who had just given the Richard Dimbleby Lecture on BBC1, in which he said that England always celebrates the arts and doesn't celebrate the sciences. In fact, he's right; there is a sort of liberal humanistic culture here which acts like art is wonderful and science is something that people should just get on with, and tell us when you've come up with a new washing machine or something. He gave the impression in his lecture that there was therefore a much better understanding of the arts than of the sciences, and I said I felt exactly the reverse was true, that people had a very poor understanding of the arts, and the reason they could happily waffle on about it was because their waffle was unchallengeable. There's such a poor conversation about it that you can say whatever crap you want to, and nobody's going to call you on it. The other thing is that everybody recognizes the power of science. We recognize the power of cloning technologies, of nuclear weapons, and so on. Everybody knows that science is powerful and could be dangerous; therefore, there's a whole lot of criticism on that basis. What people don't realize is that culture is powerful and could be dangerous, too. As long as culture is talked about as though it's a kind of nice little add-on to make things look a bit better in this sort of brutal life we all lead, as long as it's just seen as the icing on the cake, then people won't realize that it's the medium in which we're immersed, and which is forming us, which is making us what we are and what we think.

Dawkins wrote back saying my letter came at a good time, because what he's thinking about more and more is memes, rather than genes, and of course memes are what culture is about. Culture is the landscape of memes.

EDGE: Where do you see yourself going with these ideas?

ENO: One of the understandings I look for is anything that starts to take seriously the culture that ordinary people make. I find this in books such as *A Pattern Language*, by Christopher Alexander, and *How Buildings Learn*, by Stewart Brand. It's important to seek to dignify and take seriously what people who don't consider themselves experts and professionals do with their time. I would want to see the same thing done culturally, that we start to recognize that people are cultural beings. They can't help themselves. It's not a question of making a decision to become an artist. You can't help yourself, to some extent. That's an important psychological step, because it says to people: You do it.

There's another level at which I would like to say that much more profoundly; it's something I didn't talk about at all because it's a difficult issue to explain. What is cultural value and how does that come about? Nearly all of art history is about trying to identify the source of value in cultural objects. Color theories and dimension theories, golden means, all those sort of ideas, assume that some objects are intrinsically more beautiful and meaningful than others. New cultural thinking isn't like that. It says that *we* confer value on things. *We* create the value in things. It's the act of conferring that makes things valuable. Now this is very important, because so many, in fact all fundamentalist ideas, rest on the assumption that some things have intrinsic value and resonance and meaning. All pragmatists work from another assumption: No, it's us. It's us who make those meanings.

Culture is a way of getting people to that point of understanding. The work of a lot of modern culture is to say to people: You're making value. When Marcel Duchamp exhibited a lavatory, in what he called an act of deliberate aesthetic indifference, what he was saying was, Look, I can put anything in an art gallery, and I can get you to engage with that thing in a way which makes it valuable. He was quite clearly saying that it's the transaction between you and it, and this context, which creates the value.

This is something that anyone who deals with world finances would probably understand; value is conferred and the result of a system of confidences among people. But it is not something that religions generally understand. It is certainly not something that fundamentalists understand. For me, so many of the really critical bottleneck-type problems of our time come from that difficulty of understanding that it's humans who make the value in things. It didn't get there, it wasn't in there, it isn't there all the time, it wasn't made by somebody else and left there for us to find it. We made it. We put it there.

The engagement with culture is a way of understanding that. Of course, art history of the past has always used it to buttress that old idea—ah, yes, Michelangelo's *Pietà* is beautiful because these proportions have some kind of divine golden-mean-type resonance, and it communicates through to us—that the value is in the thing and we're like a radio receiver. That transmitter-receiver model is an old picture which I don't accept anymore. The value is in the transaction. The object itself can be almost irrelevant—as was Duchamp's lavatory. He could have chosen a spade, or a bicycle wheel, in fact. What he did was create the situation where he said, Here, viewer, come in and make some value. And a lot of 20th century art has been about that—about reminding us that we make things valuable, that they don't preexist in a valuable state.

5.

We Are as Gods and Have to Get Good at It

Stewart Brand

Cofounder and cochairman, The Long Now Foundation; author, Whole Earth Discipline

About forty years ago, I wore a button that said, "Why haven't we seen a photograph of the whole Earth yet?" Then we finally saw the pictures. What did it do for us?

The shift that has happened mainly has to do with climate change. Forty years ago, I could say, in the *Whole Earth Catalog*, "We are as gods, we might as well get good at it." Photographs of Earth from space had that godlike perspective.

What I'm saying now is we are as gods and have to get good at it. Necessity comes from climate change, potentially disastrous for civilization. The planet will be okay, life will be okay. We will lose vast quantities of species, probably lose the rain forests if the climate keeps heating up. So it's a global issue, a global phenomenon. It doesn't happen in just one area. The planetary perspective now is not just aesthetic. It's not just perspective. It's actually a world-sized problem that will take world-sized solutions that involve forms of governance we don't have yet. It involves technologies we are just glimpsing. It involves what ecologists call ecosystem engineering. Beavers do it, earthworms do it. They don't usually do it on a planetary scale. We have to do it on a planetary scale. A lot of sentiments and aesthetics of the environmental movement stand in the way of that.

On the other hand, a lot of the experience in the environmental movement of doing some things right, like being worried about climate very early, is part and parcel of what it will take to fix the problem. But it's going to take engineering. And environmentalists don't like engineering. It's going to take a lot more science that environmentalists have to learn to be nonselective about. They like climate science but they are not interested in nuclear science. They like climate science but they are not interested in genetic-engineering science. That's what needs to change.

But there are a couple of things the environmental movement may not change. It feels as though it's on a success roll. At least people are believing them about climate and some other things. Everything is supposed to be green. For me, this is a kind of repeat of what happened in the 1970s. We were supposed to care about solar and wind and all these good things, which was great.

Environmentalists may not change. In a way, it's almost irrelevant whether environmentalists change and take the lead, because the political situation suggests that we don't have the global institutions to deal with a global problem. A lot of the best information comes from the United Nations; the studies on cities and slums and population and things like that are about the best source you can find.

If the U.N. weren't there, we wouldn't have good information about those things, but it does not have administrative power. We probably don't want a planetary government, but I suspect that if we go too far down in a climate catastrophe, that could be something you wind up with, a bad planetary government. What you want is planetary collaboration primarily of the main governments that exist now.

This is also a flip for environmentalists and many others. From Ronald Reagan on the right and from Earth Day on the left, governments have been in ill-repute. But it's governments that run infrastructure. They run the energy systems. If European nations, North American nations, India, and China all do their governmental job of making coal illegal, make combustion the last source of energy, and adjust their economies accordingly, and they are the ones that can do that, then we might have a shot. They can do it collaboratively. Collaboratively would be much better. They need to take science more seriously than they do. It's one of the reasons China has the potential of taking the lead in this area, because all of its leadership has basically an engineering background, whereas our leadership has mostly a legal background. Europe is mixed. In India, it remains to be seen. It could go either way.

Those nations are the big makers of greenhouse gases. Those are the ones where prosperity is bringing people up the so-called energy ladder to wanting more and more grid electricity. Grid electricity right now is made with coal, which makes greenhouse gases, which is scorching the planet. They should flip to nuclear. We should be doing serious exploration of solar in space rather than on the ground where it tears up the ground because sunlight is so diluted on the surface. We will see what needs to happen over the next few years, and then it's the question of do you believe that these governments and people in general will rise to the occasion. It may take some serious catastrophes before they take it seriously.

The main thing I have changed my mind about is nuclear power. I basically went along with borrowed thinking that nuclear is bad because it involves this long-term waste-storage problem and therefore we should do other stuff. I didn't think through what the other stuff was, and the other stuff is basically about burning coal in vast quantities and putting vast quantities of greenhouse gases in the atmosphere. By opposing nuclear, I was part of the problem.

I came to a realization that the way we think about waste is bonkers. We talk about how it needs to be completely sequestered for 10,000 years, as if humans 10,000 years from now will be exactly the way we are now. One thing we can be sure of is that we won't be the way we are now even 100 or 200 years from now. We have learned from Chernobyl that when you have a totally disastrous accident with a nuclear reactor, the wildlife absolutely love it, because it chases the people away. We now have the best wildlife preserve in Europe where we had the worst nuclear accident in history, and that is interesting. This, plus various other things, added up to a flip from *nuclear is bad* to *nuclear is good* for me. The easy way to think about it is: Climate change means turning off coal, which means turning on nuclear.

This will have to be sold locally. We will have to deal with the NIMBY ("not in my backyard") problem. For my part, I would be fine with a nuclear plant in my hometown of Sausalito, California. I would be fine with nuclear-waste storage in Sausalito. The places we store nuclear waste now are in these dry casks out back of the parking lot at the hundreds of reactors in the U.S. and around the world. Waste storage is really not that big of an issue. France completely finessed it. If you want to do underground storage, there are lots of good places for it. When you look at the details on it and just look at the practicalities that emerged from what is still not a very mature industry but has great prospects for microreactors and things like that, then things that look like absolutely showstopping problems are just mild engineering problems that you solve in an engineering way and go on to deal with more interesting stuff.

I have done a considerable amount of public speaking on this subject and my talk about nuclear energy comes in two forms, the short form and the long form. The short form is when I am on a panel and I say nuclear looks pretty good. That will get a boo from an audience and raised hackles from other people on the panel. People have stacked up a bunch of arguments against nuclear: the capital costs, the waste issue, the proliferation of weapons issue. They say, well, when all these things add up you can't do nuclear. But when you take the points one by one and work through all the arguments—radioactivity is not all that serious, etc., etc.—then people find it actually a comfort in many cases just to shrug and say, okay, now that I understand the facts, nuclear doesn't look that bad. So it takes a long form. Unless people want to say, if so-and-so likes nuclear, I like nuclear, which is not a good reason to do it, then they need to go through the argument.

Another interesting issue that comes up when I talk to audiences is concern about the vast amount of energy that global computation is consuming. To the extent that running the Web takes vast quantities of energy to handle all of that cloud computing, we clearly need clean sources of that energy. Maybe some of this stuff should be off-planet. Here is a sequence that I see as plausible, just barely plausible. The major expense of working in space is the cost of getting into orbit, but we now have the possibility of diverging asteroids that would otherwise crash into earth and cause serious problems. We are looking at at least a 20 percent chance of a very serious asteroid strike in this century. We now have the capability of going out, tagging the asteroids that might be an issue, and moving them, first by ramming them and then by adjusting their orbit with what is called a gravity tractor. So that is worth doing and it is relatively inexpensive to do.

It could be any spacefaring nation, including us, that does this. There are programs that go out and look at asteroids anyway. While we are looking at them, we might as well put transponders on them, see if they are actually a threat, and move them.

So far, the private companies do not have the scale of, say, the Russian or U.S. or Indian or Japanese or Chinese space programs, the scale needed to do this kind of heavy lifting. But it's imaginable that they could have it. The great attraction of this idea is, if you can move an asteroid, you can basically mine an asteroid, and metal asteroids are damn near pure gold in an economic sense. They are made of fabulous quantities of extremely valuable metals.

In environmental terms, mining asteroids is a low-impact thing to do in the sense that there are no species that you worry about, there are no native tribes that you worry about. You would probably have issues of various nations claiming ownership of an asteroid. But once you start mining asteroids, then the whole economics of being in space flips from "It's very expensive to do science" to "It's actually a commercially valuable venture." Then you start being able to pay for getting stuff off the planet's surface, and a lot of the stuff you might build with in space you get from these asteroids. You don't have to get it up the gravity well. If you do that, then solar energy in space starts to look very plausible, and basically you can play out Marshall McLuhan's idea that after *Sputnik* there is no nature, only art. We would be doing a lot of art in space.

Laurence Smith's essay "Will We Decamp for the Northern Rim?" which ran on *Edge*, points to some possibilities for new ideas in governance.

He had more on-the-ground truth than people usually do about that area because it's so remote, about what is really going on with the forests, what is going on with the tundra, going on with the ice, and going on with the people and animals that live with the ice. One way or another, the Northern Rim is now essential. Because the ice is melting, the Northwest Passage is opening up. The northern sea route above all of Russia is opening up, and that totally changes the shipping and therefore the economic layout of the world. There was a reason that people kept looking for a Northwest Passage frantically for all those centuries, and now we have found it. It's a cruel way to do it. You just melt the ice. That is already a big deal. Humans are going up there for shipping reasons.

Climate change may also be welcomed by the people who live in Greenland, especially if they can finally grow some crops. If warming continues, and the equatorial tropics basically move north and south, and it gets drier and hotter, people will move toward the poles. There is not that much land in the Southern Hemisphere. There is a lot of land in the Northern Hemisphere.

Siberia and Canada start to look in some ways like prime land. People might want to go there. It's still dark half the year. It has some disadvantages, but it's an area that is going to have more people thinking about it, getting oil out of it, and looking for uranium in it and all the rest, and possibly living there.

But the other reason the Northern Rim is essential is that it's where a lot of methane is locked up, in the permafrost, in the form of methane clathrates. If it melts, it will release methane, a greenhouse gas twenty times more potent than carbon dioxide. It doesn't stay in the atmosphere as long but it is very potent. It can lead to big sudden jumps in greenhouse gases and in warming, and how do you reverse that? So the Northern Rim is interestingly a global domain, but there are only five or six nations that are directly involved in it.

So it may actually be a tractable world policy to do some things applied to that area, applied just to the permafrost, applied just to the ice, applied maybe just to the stratosphere above the poles. You put sulfur dioxide in the stratosphere, and you get sulfate, which thins the air, as we found with Pinatubo, the volcano that blew some years back. You put a lot of sulfur dioxide in the atmosphere, it cools everything off. You could do that just in the north. Again, there are just a few nations involved. That would be a practical way to see if that kind of human engineering might work. Basically, I think geoengineering is now in the process of gradually but steadily moving from a total taboo to the point where we have to think seriously about it; we have to do serious research and take these ideas away from the scientists and turn them over to engineers and try some stuff.

Solar has a lot going for it. I've been using solar electricity for thirty years. We used to run electric fences to keep cattle out and open gates automatically. We have a solar water heater that heats the pools. You don't have to use propane. Solar is really good on the individual building level.

On the industry level, the problem is that sunlight is so diluted. Wind has a similar issue in that the footprint of collecting it is huge. There are, I believe, about eighty projects for solar that want to go forward in Southern California that would use 1,000 square miles. Well, that 1,000 square miles is not bare land. It's living desert. It's serious natural landscape that a lot of people care about and have worked very hard to protect.

When you put in solar on an industrial scale, you basically bulldoze the landscape and make it perfectly level. It is an intensely damaging form of technology to the landscape. The same applies to wind. Though it's not as damaging to the surface of the ground, a wind farm has to be in an area that has lots of wind, and then it just completely covers that area and about the only other thing you can do with it is run cattle.

That's okay. But still, vast quantities of land are utilized for not so much energy. Both technologies so far do not have a way to store the energy that they collect when the sun is shining and the wind is blowing. So they don't actually feed into base load, i.e., that power that is always on and makes cities possible.

The downside of solar, the downside of wind, is gradually becoming apparent to environmentalists and everybody else, and we are getting used to the idea of trade-offs. A little more sophistication about trade-offs is emerging from doing actual stuff, from actually trying these things. Wind farms have the same cost problems that nuclear reactors do. They are very expensive at the beginning. Sometimes the initial investor can't stay with it. You wind up, three investors later, gradually making money. That's the norm. I don't think we understand infrastructure. I don't think we understand the economics of infrastructure. We just have big plans. We build stuff and it works out or it doesn't; we pay for it or it doesn't. We're going through a lot of that, and we will find that solar and wind are important but not important enough. That's why we are going to need to go further, certainly pushing ahead on efficiency and all that good stuff, but also finding some other energy sources that are not combustion and that are not so land-intensive as solar and wind.

Because biotechnology is probably moving as fast and as consequentially as any technology we have now, if you have an energy problem, one of the first things you do is try to sort out if there is a biotech set of solutions, some engineering approach, whether it's microbes or whatever, that will make it productive. We did the first round of biofuels, converting food into fuel. That turned out not to work so well, but it was worth trying.

Biotech fuel will definitely be a player. It could get to the point where you fulfill George Church's vision of plants that generate pipeline-ready fuels of various quality for exactly what you want.

If you want jet fuel, it will come out of the plants like maple syrup. We're a ways from that, but probably not as far as we think. It is, I suspect, not possibly the total solution, because again it's a form of solar technology.

So far all the major energy-producing biotechnologies are based on one form or another of photosynthesis. That involves making use of the very dilute sunlight that gets to the earth's surface. There are some who want to take coal and turn it into something more useful. Craig Venter wants to convert coal down in the ground into methane, which is more efficient and less harmful as a gas source than pure coal. That might work. The general understanding is that we have to go ahead and do absolutely everything, including things that we are uncertain about, because not all of them are going to work out and we need a lot of things to work out. No one thing is going to fix the problem. But we're a long way from what the people really want, which is what Bob Metcalfe calls "squanderable clean energy." We would like to have enough energy not to worry about it. I love the term. It is just such a violation of every fond notion we have that we must live frugally and that if you don't live frugally you are not living properly and all of that.

Personally, I don't think squanderable energy would be a great thing, but I love having it out there as a carrot for those who would like such a thing. Clean squanderable energy. Good—build it.

My agenda for America's climate policy? I suppose it would be energy, agriculture, climate, and urban policy. Throw in education, what the hell. Cities are green. Push cities both here and in the rest of the world. Nuclear is green. Push nuclear here and everywhere. Genetic engineering for food crops, for medical crops, for all kinds of things is green. Push that everywhere. Cohere them around the idea that we are going to have a climate policy for this country that is a model for anyone else who wants to do the same, with the hope that everybody does do the same, and we can come out of the century in at least as good of shape as we were coming into it.

Cities are green. Cities are green because they are energy-efficient. It doesn't take as much electricity to heat and cool people in the city. They use much less energy moving around. They can take mass transit because you have mass humanity. Therefore, mass humanity makes sense. It doesn't make sense living way the hell out in the boondocks. It turns out people like to live in urban areas.

So all over the world, millions of people every month are swarming into cities. They are putting up with slum conditions because it is such an improvement in terms of opportunity over the situation they had in the country. The world is losing peasant culture permanently, and the rise of the West is over because all the growing cities are elsewhere in the world: the Global South, as they say. These are the young people of the world, the innovative people of the world, the ones who take cell phones and turn them into cash systems more quickly than we can, with better cell phone signals than we have in the North.

We now have over half the world living in cities. In terms of footprint, this uses about 2.8 percent of the ice-free land area of the world. Pretty soon you will have 70 to 80 percent of the world living in cities, which by then might take up 3 percent of the land area. People move out of the country, and they stop burning wood for stoves and heating and so on. The woods come back. They stop killing the animals for meat. The animals come back. They stop drawing water out of the ground, and the aquifers come back. All of this happens as a by-product of moving away from the country and moving into the city. So cities are pretty green.

People will congregate in cities not because anybody told them to, but because they wanted to. I saw a version of this in reverse in the 1960s, when part of the notion of the time was that cities were bad, country was good. People went out to communes in the country and expired of boredom. Then they bounced back to the cities and started all these businesses, and that's our generation. We loved country life and then got over it.

Another problem we are facing is what I call the harm issue. A peculiarity of the environmental movement is that it got trapped into association with leftist politics or at least liberal politics. It's had various bad effects. Mainly it means that a whole lot of conservatives can't accept the idea that climate change is actually happening because to do that would involve them admitting that Al Gore was right about something and they can't stand the thought. That's the ridiculous situation we have gotten into. But if you start looking at the benefits and harms done by various environmental programs, just in terms of whether they are good or bad for the people, the Green Revolution was good for the masses. It staved off famine. It staved off war. For some reason, environmentalists like to think the Green Revolution was sort of a bad thing. They're wrong about that.

What Norman Borlaug (the father of the Green Revolution) and others brought to what we now call the developing world was basically Western agricultural practices that involved fertilizers, that involved herbicides and pesticides, that involved better strains of food crops created through breeding. (Now they are created by genetic engineering.) Massive quantities of food became available in parts of the world that had not had that kind of productivity in their agriculture ever before, and so it changed the world in a good way.

To look down on the Green Revolution as a bad thing strikes me as a real contradiction of liberal principles. It gets worse, though. There is a superstition about transgenic engineering—i.e., genetic engineering moving a gene from one species into another—that says it is somehow against "nature." It's not. It's the commonest form of how genes get around in nature. But if all you understand is Darwinian agricultural breeding, then you think there is something wrong with it.

Europeans decided that genetic engineering, genetically modified organisms in food, was a bad thing—a really, really bad thing and you mustn't do it. It became one of those things like animal rights or abortion, where it's okay to threaten and burn scientists and their research projects because it is so evil, so Frankensteinian. Franken-food.

This has hurt Africa, which relies on markets in Europe in order to develop cash crops, for reasons I'm still not clear about. Africa is the agricultural area that has the most need of new technology, especially better crops, which you can get faster with genetic engineering than in any other way. But of the African nations, only South Africa—which has enough of its own scientists to know when environmentalists are saying something useful and when they are not—resisted the European line.

Most of Africa bought the environmentalists' story that genetically engineered crops are somehow poison, are somehow a bad thing. Starvation is a direct result, and the neediest part of the world getting the agriculture it requires has been delayed by maybe twenty or thirty years by environmentalists. That is hard, and it is a harm to the People, with a capital P. Environmentalists have sinned.

All of the environmental organizations that I know of are opposed to genetic engineering and food crops, some quite avidly, especially those based in Europe. The American organizations are at best mildly opposed. Nobody is really neutral, that I know of, and nobody is strongly in favor, as these groups should be.

It's a problem, because genetically engineered crops, just the ones we do have, have turned out to be both economically and ecologically beneficial. They mean higher yield, which means a smaller amount of land needs to be used for crops. They mean you don't need to put pesticide on the landscape, because you have it in the plant itself. They mean that if you have the herbicide resistance in the crop, like Roundup-ready soybeans, then you don't have to put vast quantities of herbicide out there, because you can do it right at the point in the crops when the weeds are coming up, kill all the weeds, and you're ahead of the game.

So all of this is at least environmentally benign and in many cases beneficial. This is a case where the science is really, really clear, but because it goes against some superstitious, sentimental notions, the science is ignored. It means that the poor people receiving insufficient food are left out on purpose by individuals and organizations that have an irrational aversion to genetically engineered food crops. At a conference in Germany, Craig Venter was asked, "Aren't you playing God?" His response: "We're not playing."

You have to flip it back. Come on, guys. We're playing God all the time. What else is new? We are as gods and have to get good at it.

It's true that some of the biotech people now refer to what they are doing as playing nature, i.e., it's biointegrated. It's taking natural processes and conjuring with them. So what else is new? It's just natural processes that are unfamiliar to people, like how microbes make their living. What microbes do all day is what we are now doing in the labs. It's not that big a deal if you know what microbes do, and it's a huge deal if you don't know what microbes do.

One of the peculiarities of this planet's atmosphere is that it's run by microbes, and we are only gradually figuring out how that actually works, what role the oceans especially have in it—ocean microbial ecology in chemical transfer and carbon dioxide management and carbon sequestration and all of that. Plus, now that we have gotten down to the nano scale with our technology, with things like metagenomic shotgun-sequencing of whole environments of microbes, we are starting to understand their life and how they run the world. So I guess my solution to our current problems is, if you have a question you're not sure how to answer, ask a microbe.

6.

Turing's Cathedral

A Visit to Google on the Occasion of the 60th Anniversary of John von Neumann's Proposal for a Digital Computer

George Dyson

Science historian; author, Darwin Among the Machines *and* Project Orion

In the digital universe, there are two kinds of bits: bits that represent structure (differences in space) and bits that represent sequence (differences in time). Digital computers—as formalized by Alan Turing, and delivered by John von Neumann—are devices that translate between these two species of bits according to definite rules.

On October 24, 1945, at the Institute for Advanced Study in Princeton, New Jersey, mathematician John von Neumann began seeking funding to build a machine that would do this at electronic speeds. "I am sure that the projected device, or rather the species of devices of which it is to be the first representative, is so radically new that many of its uses will become clear only after it has been put into operation," he wrote to Lewis Strauss. "Uses which are likely to be the most important are by definition those which we do not recognize at present because they are farthest removed from our present sphere."

Von Neumann received immediate support from the Army, the Navy, and the Air Force, but the main sponsor soon became the

United States Atomic Energy Commission, or AEC. This deal with the devil was hard to resist. "The Army contract provides for general supervision by the Ballistic Research Laboratory of the Army, whereas the AEC provides for supervision by von Neumann," the Institute administration explained in 1949. When the machine finally became operational in 1951, it had 5 kilobytes of random-access memory: a $32 \times 32 \times 40$ matrix of binary digits, stored as a flickering pattern of electrical charge, shifting from millisecond to millisecond on the surface of 40 cathode-ray tubes.

The codes that inoculated this empty universe were based upon the architectural principle that a pair of 5-bit coordinates could uniquely identify a memory location containing a string of 40 bits. These 40 bits could include not only data (numbers that mean things) but executable instructions (numbers that do things)—including instructions to transfer control to another location and do something else.

By breaking the distinction between numbers that *mean* things and numbers that *do* things, von Neumann unleashed the power of the stored-program computer, and our universe would never be the same. It was no coincidence that the chain reaction of addresses and instructions within the core of the computer resembled a chain reaction within the core of an atomic bomb. The driving force behind the von Neumann project was the push to run large-scale Monte Carlo simulations of how the implosion of a subcritical mass of fissionable material could lead the resulting critical assembly to explode.

The success of Monte Carlo led to compact, predictable fission explosives, and this, coupled with more Monte Carlo (and more Stan Ulam) led to the "Super," or hydrogen bomb. But the actual explosion of digital computing has overshadowed the threatened explosion of the bombs. From an initial nucleus of 4×10^4 bits changing state at kilocycle speed, von Neumann's archetype has proliferated to individual matrices of more than 10^9 bits, running at speeds of more than 10^9 cycles per second, interconnected by an extended address matrix encompassing up to 10^9 remote hosts. This growth continues to speed up. Over 10^{10} transistors are now produced each second, and many of these are being incorporated into devices—no longer just computers—that have an IP (Internet Protocol) address. The current 32-bit IP address space will be exhausted within 10 years or less.

In the early 1950s, when mean time between memory failure was measured in minutes, no one imagined that a system depending on every bit being in exactly the right place at exactly the right time could be scaled up by a factor of 10^{13} in size, and down by a factor of 10^6 in time. Von Neumann, who died prematurely in 1957, became increasingly interested in understanding how biology has managed (and how technology might manage) to construct reliable organisms out of unreliable parts. He believed the von Neumann architecture would soon be replaced by something else. Even if codes could be completely debugged, million-cell memories could never be counted upon, digitally, to behave consistently from one kilocycle to the next.

Fifty years later, thanks to solid-state microelectronics, the von Neumann matrix is going strong. The problem has shifted from how to achieve reliable results using sloppy hardware, to how to achieve reliable results using sloppy code. The von Neumann architecture is here to stay. But new forms of architecture, built upon the underlying layers of Turing–von Neumann machines, are starting to grow. What's next? Where was von Neumann heading when his program came to a halt?

As organisms, we possess two outstanding repositories of information: the information conveyed by our genes, and the information stored in our brains. Both of these are based upon non–von Neumann architectures, and it is no surprise that von Neumann became fascinated with these examples as he left his chairmanship of the AEC (where he had succeeded Lewis Strauss) and began to lay out the research agenda that cancer prevented him from following up. He considered the second example in his posthumously published *The Computer and the Brain*.

> The message-system used in the nervous system . . . is of an essentially *statistical* character. In other words, what matters are not the precise positions of definite markers, digits, but the statistical characteristics of their occurrence . . . a radically different system of notation from the ones we are familiar with in ordinary arithmetics and mathematics . . . Clearly, other traits of the (statistical) message could also be used: indeed, the frequency referred to is a property of a single train of pulses whereas every one of the relevant nerves consists of a large number of fibers, each of which transmits numerous trains of pulses. It is, therefore, perfectly plausible that certain (statistical) relationships between such trains of pulses should also transmit information. . . . Whatever language the central nervous system is using, it is characterized by less logical and arithmetical depth than what we are normally used to [and] must structurally be essentially different from those languages to which our common experience refers.

Or, as his friend Stan Ulam put it, "What makes you so sure that mathematical logic corresponds to the way we think?"

Pulse-frequency coding, whether in a nervous system or a probabilistic search engine, is based on statistical accounting for what connects where, and how frequently connections are made between given points. As von Neumann explained in 1948: "A new, essentially logical, theory is called for in order to understand high-complication automata and, in particular, the central nervous system. It may be, however, that in this process logic will have to undergo a pseudomorphosis to neurology to a much greater extent than the reverse."

Von Neumann died just as the revolution in molecular biology, sparked by the elucidation of the structure of DNA in 1953, began to unfold. Life as we know it is based on digitally coded instructions, translating between sequence and structure (from nucleotides to proteins) exactly as Turing prescribed. Ribosomes and other cellular machinery play the role of processors: reading, duplicating, and interpreting the sequences on the tape. But this uncanny resemblance has distracted us from the completely different method of addressing by which the instructions are carried out.

In a digital computer, the instructions are in the form of COMMAND (ADDRESS), where the address is an exact (either absolute or relative) memory location, a process that translates informally into "DO THIS with what you find HERE and go THERE with the result." Everything depends not only on precise instructions, but on HERE, THERE, and WHEN being exactly defined. It is almost incomprehensible that programs amounting to millions of lines of code, written by teams of hundreds of people, are able to go out into the computational universe and function as well as they do given that one bit in the wrong place (or at the wrong time) can bring the process to a halt.

Biology has taken a completely different approach. There is no von Neumann address matrix, just a molecular soup, and the instructions say simply "DO THIS with the next copy of THAT which comes along." The results are far more robust. There is no unforgiving central address authority, and no unforgiving central clock. This ability to take general, organized advantage of local, haphazard processes is exactly the ability that (so far) has distinguished information processing in living organisms from information processing by digital computers.

Of course, dreams of object-oriented programming languages and asynchronous processing have been around almost as long as digital computing, and content-addressable memory was one of the alternative architectures that Julian Bigelow, von Neumann's original chief architect, and many others had long had in mind. As Bigelow explained in 1965:

> For man-made electronic computers, a practice adopted, whereby
> events are represented with serial dependence in time, has resulted
> in computing apparatus that must be built of elements that are,
> to a large extent, strictly independent across space-dimensions.
> Accomplishment of the desired time-sequential process on a given
> computing apparatus turns out to be largely a matter of specifying
> sequences of addresses of items which are to interact . . . With
> regard to the explicit address nuisance, studies have been made of
> the possibility of causing various elementary pieces of information
> situated in the cells of a large array (say, of memory) to enter into
> a computation process without explicitly generating a coordinate
> address in "machine-space" for selecting them out of the array.

Hardware-based content-addressable memory is used, on a local scale, in certain dedicated high-speed network routers, but template-based addressing did not catch on widely until Google (and brethren) came along. Google is building a new, content-addressable layer overlying the von Neumann matrix underneath. The details are mysterious but the principle is simple: It's a map. And, as Dutch (and other) merchants learned in the 16th century, great wealth can be amassed by Keepers of the Map.

We call this a search engine—a content-addressable layer that makes it easier for *us* to find things, share ideas, and retrace our steps. That's a big leap forward, but it isn't a universe-shifting revolution equivalent to von Neumann's breaking the distinction between numbers that *mean* things and numbers that *do* things in 1945.

However, once the digital universe is thoroughly mapped, and initialized by us searching for meaningful things and following meaningful paths, it will inevitably be colonized by codes that will start *doing* things with the results. Once a system of template-based addressing is in place, the door is opened to code that can interact directly with other code, free at last from a rigid bureaucracy requiring that every bit be assigned an exact address. You can write instructions (and a few people already are) that say, "Do THIS with THAT"— without having to specify exactly WHERE or WHEN. This revolution will start with simple, basic coded objects, on the level of nucleotides heading out on their own and bringing amino acids back to a collective nest. It is 1945 all over again.

And it is back to Turing, who in his 1948 report to the National Physical Laboratory on intelligent machinery advised that "intellectual activity consists mainly of various kinds of search." It was Turing, in 1936, who showed von Neumann that digital computers are able to solve most—but not all—problems that can be stated in finite, unambiguous terms. They may, however, take a very long time to produce an answer (in which case you build faster computers), or it may take a very long time to ask the question (in which case you hire more programmers). Computers have been getting better and better at providing answers—but only to questions that programmers are able to ask.

We can divide the computational universe into three sectors:

computable problems; noncomputable problems (that can be given a finite, exact description but have no effective procedure to deliver a definite result); and, finally, questions whose answers are, in principle, computable, but that, in practice, we are unable to ask in unambiguous language that computers can understand.

We do most of our computing in the first sector, but we do most of our living (and thinking) in the third. In the real world, most of the time, finding an answer is easier than defining the question. It's easier to draw something that looks like a cat, for instance, than to describe what, exactly, makes something look like a cat. A child scribbles indiscriminately, and eventually something appears that resembles a cat. A solution finds the problem, not the other way around. The world starts making sense, and the meaningless scribbles (and a huge number of neurons) are left behind.

This is why Google works so well. All the answers in the known universe are there, and some very ingenious algorithms are in place to map them to questions that people ask.

"An argument in favor of building a machine with initial randomness is that, if it is large enough, it will contain every network that will ever be required," advised Turing's former assistant and cryptanalyst Irving J. Good, speaking at IBM in 1958. A network, whether of neurons, computers, words, or ideas, contains solutions, waiting to be discovered, to problems that need not be explicitly defined. It is much easier to find explicit answers than to ask explicit questions. And some will be answers to questions that programmers will never have to ask.

"The whole human memory can be, and probably in a short time will be, made accessible to every individual," wrote H. G. Wells in his 1938 prophecy *World Brain*. "This new all-human cerebrum need not be concentrated in any one single place. It can be reproduced exactly and fully, in Peru, China, Iceland, Central Africa, or wherever else seems to afford an insurance against danger and interruption. It can have at once, the concentration of a craniate animal and the diffused vitality of an amoeba." Wells foresaw not only the distributed intelligence of the World Wide Web, but the inevitability that this intelligence would coalesce, and that power, as well as knowledge, would fall under its domain. "In a universal organization and clarification of knowledge and ideas . . . in the evocation, that is, of what I have here called a World Brain . . . in that and in that alone, it is maintained, is there any clear hope of a really Competent Receiver for world affairs . . . We do not want dictators, we do not want oligarchic parties or class rule, we want a widespread world intelligence conscious of itself."

My visit to Google? Despite the whimsical furniture and other toys, I felt I was entering a 14th-century cathedral—not in the 14th century, but in the 12th century, while it was being built. Everyone was busy carving one stone here and another stone there, with some invisible architect getting everything to fit. The mood was playful, yet there was a palpable reverence in the air. "We are not scanning all those books to be read by people," explained one of my hosts after my talk. "We are scanning them to be read by an AI."

I found myself recollecting the words of Alan Turing, in his seminal paper *Computing Machinery and Intelligence*, a founding document in the quest for true AI. "In attempting to construct such machines we should not be irreverently usurping His power of creating souls, any more than we are in the procreation of children," Turing had advised. "Rather we are, in either case, instruments of His will providing mansions for the souls that He creates."

Google is Turing's cathedral, awaiting its soul. We hope. In the words of an unusually perceptive friend: "When I was there, just before the IPO, I thought the coziness to be almost overwhelming. Happy golden retrievers running in slow motion through water sprinklers on the lawn. People waving and smiling, toys everywhere. I immediately suspected that unimaginable evil was happening somewhere in the dark corners. If the devil would come to earth, what place would be better to hide?"

For thirty years I have been wondering, What indication of its existence might we expect from a true AI? Certainly not any explicit revelation, which might spark a movement to pull the plug. Anomalous accumulation or creation of wealth might be a sign, or an unquenchable thirst for raw information, storage space, and processing cycles, or a concerted attempt to secure an uninterrupted, autonomous power supply. But the real sign, I suspect, would be a circle of cheerful, contented, intellectually and physically well-nourished people surrounding the AI. There wouldn't be any need for True Believers, or the downloading of human brains or anything sinister like that: just a gradual, gentle, pervasive and mutually beneficial contact between us and a growing something else. This remains a non-testable hypothesis, for now. The best description comes from science-fiction writer Simon Ings: "When our machines overtook us, too complex and efficient for us to control, they did it so fast and so smoothly and so *usefully*, only a fool or a prophet would have dared complain."

7.

Time to Start Taking the Internet Seriously

David Gelernter

Professor of computer science, Yale University; chief scientist, Mirror Worlds Technologies; author, Mirror Worlds

1. No moment in technology history has ever been more exciting or dangerous than now. The Internet is like a new computer running a flashy, exciting demo. We have been entranced by this demo for fifteen years. But now it is time to get to work, and make the Internet do what we want it to.

2. One symptom of current problems is the fundamental puzzle of the Internet. (Algebra and calculus have fundamental theorems; the Internet has a fundamental puzzle.) *If this is the information age, what are we so well-informed about?* What do our children know that our parents didn't? (Yes, they know how to work their computers, but that's easy compared to—say—driving a car.) I'll return to this puzzle.

3. Here is a simpler puzzle, with an obvious solution. Wherever computers exist, nearly everyone who writes uses a word processor. The word processor is one of history's most successful inventions. Most people call it not just useful but indispensable. Granted that the word processor is indeed indis-

pensable, what good has it done? We say we can't do without it; but if we had to give it up, what difference would it make? Have word processors improved the quality of modern writing? What has the indispensable word processor accomplished?

4. It has increased not the quality but the quantity of our writing—"our" meaning society's as a whole. The Internet for its part has increased not the quality but the quantity of the information we see. Increasing quantity is easier than improving quality. Instead of letting the Internet solve the easy problems, it's time we got it to solve the important ones.

5. Consider Web search, for example. Modern search engines combine the functions of libraries and business directories on a global scale, in a flash: a lightning bolt of brilliant engineering. These search engines are indispensable—just like word processors. But they solve an easy problem. It has always been harder to find the right person than the right fact. Human experience and expertise are the most valuable resources on the Internet—if we could find them. Using a search engine to find (or be found by) the right person is a harder, more subtle problem than ordinary Internet search. Small pieces of the problem have been attacked; in the future we will solve this hard problem in general, instead of being satisfied with windfalls and the lowest-hanging fruit on the technology tree.

6. We know that the Internet creates "information overload," a problem with two parts: increasing number of information sources and increasing in-

formation flow per source. The first part is harder: It's more difficult to understand five people speaking simultaneously than one person talking fast—especially if you can tell the one person to stop temporarily, or go back and repeat. Integrating multiple information sources is crucial to solving information overload. Blogs and other anthology-sites integrate information from many sources. But we won't be able to solve the overload problem until each Internet user can choose for himself what sources to integrate, and can add to this mix the most important source of all: his own personal information—his e-mail and other messages, reminders and documents of all sorts. To accomplish this, we merely need to turn the whole Cybersphere on its side, so that time instead of space is the main axis.

7. In the last paragraph I wrote "each Internet user"; but users of any computing system ought to have a simple, uniform operating system and interface. Users of the Internet still don't.

8. Practical business: Who will win the tug-of-war between private machines and the Cloud? Will you store your personal information on your own personal machines, or on nameless servers far away in the Cloud, or both? Answer: in the Cloud. The Cloud (or the Internet Operating System, IOS—"Cloud 1.0") will take charge of your personal machines. It will move the information you need at any given moment onto your own cell phone, laptop, pad, pod—but will always keep charge of the master copy. When you make changes to any document, the

changes will be reflected immediately in the Cloud. Many parts of this service are available already.

9. Because your information will live in the Cloud and only make quick visits to your personal machines, all your machines will share the same information automatically; a new machine will be useful the instant you switch it on; a lost or stolen machine won't matter—the information it contains will evaporate instantly. The Cloud will take care that your information is safely encrypted, distributed, and secure.

10. Practical business: Small computers have been the center of attention lately, and this has been the decade of the cell phone. Small devices will continue to thrive, but one of the most important new developments in equipment will be at the other end of the size spectrum. In offices and at home, people will increasingly abandon conventional desktop and laptop machines for large-screen computers. You will sit perhaps seven feet away from the screen, in a comfortable chair, with the keyboard and controls in your lap. Work will be easier and eyestrain (this is important) will decrease. Large-screen computers will change the shape of office buildings and create their own new architecture. Office workers will spend much of their time in large-screen computer modules that are smaller than most private offices today, but more comfortable. A building designed around large-screen computers might have modules (for example) stacked in many levels around a central court; the column whose walls consist of stacked modules might spiral helically as it rises . . .

David Gelernter

11. The Internet will never create a new economy based on voluntary instead of paid work—but it can help create the best economy in history, where new markets (a free market in education, for example) change the world. Good news!—the Net will destroy the university as we know it (except for a few unusually prestigious or beautiful campuses). The Net will never become a mind, but it can help us change our ways of thinking and change, for the better, the spirit of the age. This moment is also dangerous: Virtual universities are good, but virtual nations, for example, are not. Virtual nations—whose members can live anywhere, united by the Internet—threaten to shatter mankind like glass into razor-sharp fragments that draw blood. We know what virtual nations can be like: Al-Qaeda is one of the first.

12. In short: It's time to think about the Internet instead of just letting it happen.

13. The traditional website is static, but the Internet specializes in flowing, changing information. The "velocity of information" is important—not just the facts but their rate and direction of flow. Today's typical website is like a stained-glass window, many small panels leaded together. There is no good way to change stained glass, and no one expects it to change. So it's not surprising that the Internet is now being overtaken by a different kind of cyberstructure.

14. The structure called a cyberstream or lifestream is better suited to the Internet than a conventional website because it shows information-in-motion, a

rushing flow of fresh information instead of a stagnant pool.

15. Every month, more and more information surges through the Cybersphere in lifestreams—some called blogs, feeds, activity streams, event streams, Twitter streams. All these streams are specialized examples of the cyberstructure we called a lifestream in the mid–1990s: a stream made of all sorts of digital documents, arranged by time of creation or arrival, changing in real time; a stream you can focus and thus turn into a different stream; a stream with a past, present, and future. The future flows through the present into the past at the speed of time.

16. Your own information (all your communications, documents, photos, videos), including cross-network information (phone calls, voice messages, text messages), will be stored in a lifestream in the Cloud.

17. There is no clear way to blend two standard websites together, but it's obvious how to blend two streams. You simply shuffle them together like two decks of cards, maintaining time order—putting the earlier document first. Blending is important because we must be able to add and subtract in the Cybersphere. We add streams together by blending them. Because it's easy to blend any group of streams, it's easy to integrate stream-structured sites so we can treat the group as a unit, not as many separate points of activity; and integration is important to solving the information-overload problem. We subtract streams by searching or focusing. Searching a stream for

David Gelernter

"snow" means that I subtract every stream-element that doesn't deal with snow. Subtracting the "not snow" stream from the mainstream yields a "snow" stream. Blending streams and searching them are the addition and subtraction of the new Cybersphere.

18. Nearly all flowing, changing information on the Internet will move through streams. You will be able to gather and blend together all the streams that interest you. Streams of world news or news about your friends, streams that describe prices or auctions or new findings in any field, or traffic, weather, markets—they will all be gathered and blended into one stream. Then your own personal lifestream will be added. The result is your mainstream: different from all others; a fast-moving river of all the digital information you care about.

19. You can turn a knob and slow down your mainstream: less-important stream-elements will flow past invisibly and won't distract you, but will remain in the stream and appear when you search for them. You can rewind your lifestream and review the past. If an important-looking document or message sails past and you have no time to deal with it now, you can copy the document or message into the future (copy it to "this evening at ten," say); when the future arrives, the document appears again. You can turn a different knob to make your fast-flowing stream spread out into several slower streams, if you have space enough on your screen to watch them all. And you can gather those separate streams back together whenever you like.

20. Sometimes you will want to listen to your stream instead of watching it (perhaps while you're driving, or sitting through a boring meeting or lecture). Software will read text aloud, and eventually will describe pictures, too. When you watch your high-definition TV, you might let the stream trickle down one side of the screen, so you can stay in touch with your life.

21. It's simple for the software that runs your lifestream to learn about your habits; simple to figure out which e-mails (for example) or social updates or news stories you are likely to find important and interesting. It will therefore be easy for software to highlight the stream elements you're apt to find important, and let the others rush by quickly without drawing your attention.

22. Lifestreams will make it even easier than it is today for software to learn the details of your life and predict your future actions. The potential damage to privacy is too large and important a problem to discuss here. Briefly, the question is whether the crushing blows to privacy from many sources over the past few decades will make us crumple and surrender, or fight harder to protect what remains.

23. The Internet's future is not Web 2.0 or 200.0 but the post-Web, where time instead of space is the organizing principle. Instead of many stained-glass windows, instead of information laid out in space, like vegetables at a market, the Net will be many streams of information flowing through time. The Cybersphere as a whole equals every stream in the

David Gelernter

Internet blended together: the whole world telling its own story. (But the world's own story is full of private information—and so, unfortunately, no human being is allowed to hear it.)

24. Ten years ago I wrote about the growing importance of lifestreams. Last year, the technology journalist Erick Schonfeld asked in a news story whether a certain large company "can take the central communication model of social networks—the lifestream—and pour it back into its IM clients." (The story was headlined "Bebo Zeroes in on Lifestreaming for the Masses.") Lifestreaming is a word that is now used generically, and streams are all over the Net. Ten years ago I described the computer of the future as a "scooped-out hole in the beach where information from the Cybersphere wells up like seawater." Today the spread of wireless coverage and the growing power of mobile devices means that information does indeed well up almost anywhere you switch on your laptop or cell phone; and anywhere at all will be true before long.

25. From which we learn that (a) making correct predictions about the technology future is easy, and (b) writers should remember to put their predictions in suitably poetic language, so it's easy to say they were right.

26. If we think of time as orthogonal to space, a stream-based, time-based Cybersphere is the traditional Internet flipped on its side in digital space-time. The traditional web-shaped Internet consists (in effect) of many flat panels chaotically connected. Instead

of flat sites, where information is arranged in space, we want deep sites that are slices of time. When we look at such a site onscreen, it's natural to imagine the past extending into (or beyond) the screen, and the future extending forward in front of the screen; the future flows toward the screen, into the screen, and then deeper into the space beyond the screen.

27. The Internet is no topic like cell phones or video-game platforms or artificial intelligence; it's a topic like education. It's that big. Therefore, beware: To become a teacher, master some topic you can teach; don't go to education school and master nothing. To work on the Internet, master some part of the Internet: engineering, software, computer science, communication theory, economics or business, literature or design. Don't go to Internet school and master nothing. There are brilliant, admirable people at Internet institutes. But if these institutes have the same effect on the Internet that education schools have had on education, they will be a disaster.

28. Returning to our fundamental riddle: If this is the information age, what do our children know that our parents didn't? The answer is "now." They know about *now*.

29. Internet culture is a culture of nowness. The Internet tells you what your friends are doing and the world news now; the state of the shops and markets and weather now; public opinion, trends, and fashions now. The Internet connects each of us to countless sites right now—to many different places at one moment in time.

David Gelernter

30. Nowness is one of the most important cultural phenomena of the modern age: The Western world's attention shifted gradually from the deep but narrow domain of one family or village and its history to the (broader but shallower) domains of the larger community, the nation, the world. The cult of celebrity, the importance of opinion polls, the decline in the teaching and learning of history, the uniformity of opinions and attitudes in academia and among other educated elites—they are all part of one phenomenon. Nowness ignores all other moments but this. In the ultimate Internet culture, flooded in nowness like a piazza flooded in seawater, drenched in a tropical downpour of nowness, everyone talks alike, dresses alike, thinks alike.

31. As I wrote at the start of this piece, no moment in technology history has ever been more exciting or dangerous than now. As we learn more about now, we know less about *then*. The Internet increases the supply of information hugely, but the capacity of the human mind not at all. (Some scientists talk about artificially increasing the power of minds and memories—but then they are no longer talking about human beings. They are discussing some new species we know nothing about. And in this field, we would be fools to doubt our own ignorance.) The effect of nowness resembles the effect of light pollution in large cities, which makes it impossible to see the stars. A flood of information about the present shuts out the past.

32. But—the Internet could be the most powerful device

ever invented for understanding the past, and the texture of time. Once we understand the inherent bias in an instrument, we can correct it. The Internet has a large bias in favor of now. Using lifestreams (which arrange information in time instead of space), historians can assemble, argue about, and gradually refine timelines of historical fact. Such timelines are not history, but they are the raw material of history. They will be bitterly debated and disputed—but it will be easy to compare two different versions (and the evidence that supports them) side by side. Images, videos, and text will accumulate around such streams. Eventually they will become shared cultural monuments in the Cybersphere.

33. Before long, all personal, familial, and institutional histories will take visible form in streams. A lifestream is tangible time: As life flashes past on water skis across time's ocean, a lifestream is the wake left in its trail. Dew crystallizes out of the air along cool surfaces; streams crystallize out of the Cybersphere along veins of time. As streams begin to trickle and then rush through the spring thaw in the Cybersphere, our obsession with nowness will recede, the dikes will be repaired, and we will clean up the damaged piazza of modern civilization.

34. Anyone who has ever looked through a telescope at the moon close up has seen it drift out of sight as the Earth slowly spins. In the future, the Cybersphere will drift, too: If you have investigated one topic long enough for your attention to grow slack and your mind to wander, the Net will respond by let-

ting itself drift slowly into new topics, new domains: not ones with obvious connections to the topic you've been studying, but new topics that have deep emotional connections to the previous ones, connections that will no doubt make sense only to you.

35. The Internet today is, after all, a machine for reinforcing our prejudices. The wider the selection of information, the more finicky we can be about choosing just what we like and ignoring the rest. On the Net we have the satisfaction of reading only opinions we already agree with, only facts (or alleged facts) we already know. You might read ten stories about ten different topics in a traditional newspaper; on the Net, many people spend that same amount of time reading ten stories about the same topic. But again, once we understand the inherent bias in an instrument, we can correct it. One of the hardest, most fascinating problems of this cyber-century is how to add "drift" to the Net, so that your view sometimes wanders (as your mind wanders when you're tired) into places you hadn't planned to go. Touching the machine brings the original topic back. We need help overcoming rationality sometimes, and allowing our thoughts to wander and metamorphose as they do in sleep.

36. Pushing the multimegaton jumbo jet of human thought-style backward a few inches, back in the direction of dream logic, might be the Internet's greatest accomplishment. The best is yet to be.

8.

Indirect Reciprocity, Assessment Hardwiring, and Reputation

Karl Sigmund

Professor of mathematics, University of Vienna; author,
Games of Life

Direct reciprocity is an idea that we can see every day in every household. If my wife cooks, for instance, I will always wash the dishes. Otherwise, cooperation breaks down. This is a commonplace situation, and similar exchanges have been measured and studied for many years. Martin Nowak and I began working together on this idea of direct reciprocity after meeting at an Austrian mountain retreat in the late 1980s. I was there to deliver my lecture on Robert Axelrod, whose book, *The Evolution of Cooperation*, was already a classic. It did wonders for the study of direct reciprocity, presenting a lot of his own work, and also leading to more research. At the time that Axelrod wrote his book there were already hundreds of papers by psychologists, philosophers, and mathematicians on what has been called the prisoner's dilemma. Afterward, there were thousands. It launched a big field.

The simplest instance of the prisoner's dilemma would be if, in separate rooms, you have two players, each of whom has to decide whether or not to give a gift to the other. According to the rules of the game, each gift-giver may give the other player $3, but doing so will cost him $1. Both players must make this decision at the same time. If both make the decision to give to the other player, both receive $3 while only having to pay $1, so both end up with a $2 net benefit. If both decline to give, it costs them nothing, and the payoff in both cases is zero. The rub is that if one player gives to the other player, and that other player does not at the same time decide to give in return, then one player is exploited. He has given $1 but gets nothing in return, whereas the other player, the nasty exploiter, gets $3 and is best off.

In the simplest prisoner's dilemma, the game is played only once. If there is no future interaction, it's obvious what you should do, because one only has to consider two possibilities. The other player can give a gift, in which case it's very good to accept it and not to give anything in return. You get $3 and it costs you nothing. If the other player does not give a gift, then it is again to your advantage not to give the other player anything because you would be foolish to pay him something when he doesn't return the favor. Under both circumstances, no matter what the other player does, you should defect by not giving a gift.

However, if the other player thinks the same thing, then you both end up with nothing. If you had both been generous you would both have earned $2 by the game. This means that pursuing your selfish interest in a rational way by thinking out the alternatives and doing what is best for you will actually lead to a solution that is bad for you. If you follow your instincts, which, if you don't know the other players, probably tell you to try cooperating and to be generous, then you will fare well. There is a distinction between the benefit of the group, so to speak—each of you gets $2—and a selfish benefit. If both you and the other player try hard for the selfish benefit, no one gets anything.

For the next ten years, Martin and I milked the prisoner's dilemma with pleasure and success. At the time, we were looking for other simple but interesting economic experiments and situations. Martin, for instance, introduced the spatial prisoner's dilemma. Here you don't form bonds with just anyone in a population and play two hundred rounds of the prisoner's dilemma with that person, but you interact only with your immediate neighbors. This is, of course, much more realistic, because usually you do not interact with just anybody in the town of Vienna, say, but only within a very small social network.

This study of the prisoner's dilemma went on for quite some time, and I must say I was happy when we reached an opening into what is now called indirect reciprocity about six years ago. Martin and I were the first to formalize the idea, creating a mathematical model to analyze this precisely. We wrote the first major paper that talked about how to analyze this question and then to set up experiments, and now there are dozens of groups that are actively involved.

There was one idea in particular in the literature already that was not taken very seriously. People thought the prisoner's dilemma game could be played according to a strategy called Tit-for-Tat. This strategy says that whenever you meet a new person, in the first round you should cooperate, and from then on you should do whatever the other fellow has done in the previous round. But real life is actually subtler. When you meet a new partner, you possibly know a bit about his past. You have not interacted with him, but he has interacted with other people. If you know that he defected against another person it would be good for you to start by expecting defection, and therefore to defect yourself. This has been called Observer Tit-for-Tat. It is just like Tit-for-Tat, except that in the first round you do not necessarily cooperate; you cooperate only if you know that in his interactions with third parties this person has been nice.

Let me explain this very carefully. Tit-for-Tat is the completely natural strategy. There is no inventor of this strategy, although it was submitted to prisoner's dilemma tournaments by game theorist Anatol Rapoport. He submitted the simplest of all strategies, which consisted of two lines of Fortran programming. The strategy is simply that in the first round, when you don't know the other player, you cooperate with the other player. You give him the benefit of the doubt, so to speak. From then onward, you mimic whatever the other player did in the previous round. If he was nasty then you are nasty. If he was friendly and cooperative, you are friendly and cooperative. This strategy was extremely successful in computer tournaments.

In a computer tournament, however, you can practically exclude the possibilities of mistakes and errors. In real life, it is quite likely that mistakes will occasionally happen. You can intend to do a nice thing and just happen not to have whatever it takes to do it. You can misunderstand an action of your partner. You can be in a bad mood because somebody else has hurt you that day and you stupidly get your revenge on the wrong person. All these things can happen.

If you play Tit-for-Tat and are subject to these mistakes, then you are likely to get embroiled in needless mutual punishment. If, for instance, the other player plays also Tit-for-Tat and made a mistake by defecting in the previous round, you would punish him by defecting in this round. He will punish you, in turn, by defecting in the next round. And then you will punish him again, and so on, and so on, creating an endless vendetta. This continues until the next mistake puts a stop to it, although actually the next mistake can make things even worse. It could happen that from then on you keep punishing each other in every round mutually— not alternating, but simultaneously. At this point, the situation is fairly hopeless.

Robert May suggested in an article in *Nature* that people should really use more generous strategies, allowing for occasional for-giveness. In our first paper in *Nature*, Martin and I presented a strategy called Generous Tit-for-Tat. Whenever the other player has done you a good turn, then there is a 100 percent chance that you will reply with a good action. But if the other player has done you a bad turn, then you will only return this bad action with a certain probability, depending on the ratio of cost-to-benefit—say, only with a probability of 35 percent. Therefore, there is a rather high probability that the cycle of mutual punishment which starts with an erroneous defection will be broken after one or two rounds. At that point, the players are again in a good mood and cooperate with each other until the next mistake happens. Gener-ous Tit-for-Tat was a very obvious and very robust solution to this problem of what happens when mistakes occur.

Later we found another even more robust strategy than Gener-ous Tit-for-Tat. This was later called Pavlov's strategy, a name that is not the best possible, but that has stuck. Pavlov's strategy says that you should cooperate if and only if in the previous round you and your co-player have done the same thing. According to this strategy:

- *If you both cooperated*, then you cooperate.
- *If you have both defected*, then you should also cooperate.
- *If you have cooperated and the other player has defected*, then you should defect in the next round.
- *If you defected and the other player has cooperated*, then you should again defect in the next round.

At first glance, the strategy looks bizarre, but in our computer simulation it turned out that it always won in an environment where mistakes were likely. In the end, it was almost always the dominating strategy in the population. Almost everyone was playing Pavlov's strategy, and it was very stable; it was much better than Tit-for-Tat.

Later we understood that this strategy is actually not so strange. It is the simplest learning mechanism that you can imagine. This is a win-stay/lose-shift learning mechanism that has already been studied in animals—for training horses and so on—for a hundred years. It says simply that if a person gets a bad result, then he is less likely to repeat the former move. And if a person has a good outcome and wins, then he is more likely to repeat the former move because it was successful. It is simply reward and punishment in action. If one studies this in the context of the prisoner's dilemma, one gets exactly Pavlov's strategy. For instance, if you have defected and the other player has cooperated, then according to the prisoner's dilemma, you have exploited the other player and your payoff is very high. You are very happy, and so you repeat your move, therefore defecting again in the next round. However, if you have cooperated and the other player has defected, then you have been exploited. You are very unhappy, and you are going to switch to another move. You have cooperated in the past, but now you are going to defect.

These are theoretical experiments, but students of animal behavior also did lots of real-life experiments. They even turned to humans with this series of experiments. People like Manfred Milinski, who is director of a Max Planck Institute in Germany and a very straight, very hard-nosed student of animal behavior, started a new career in studying human nature. He used students from Switzerland and Germany to set up experiments based on the prisoner's dilemma game to check whether people were likely to play this Pavlov strategy or not. He found that, indeed, there is strong evidence that Pavlov's strategy is quite widespread in humans.

These ideas fed into our work on indirect reciprocity, a concept that was first introduced by Robert Trivers in a famous paper in the 1970s. I recall that he mentioned this idea obliquely when he wrote about something he called general altruism. Here you give something back not to the person to whom you owe something, but to somebody else in society. He pointed out that this also works with regard to cooperation at a high level. Trivers didn't go into details, because at the time it was not really at the center of his thinking. He was mostly interested in animal behavior, and so far indirect reciprocity has not been proven to exist in animal behavior. It might exist in some cases, but ethologists are still debating the pros and cons.

In human societies, however, indirect reciprocity has a very striking effect. There is a famous anecdote about the American baseball player Yogi Berra, who said something to the effect of: I make a point of going to other people's funerals because otherwise they won't come to mine. This is not as nonsensical as it seems. If a colleague of the university, for instance, goes faithfully to every faculty member's funeral, then the faculty will turn out strongly at his. Others reciprocate. It works. We think instinctively in terms of direct reciprocation—when I do something for you, you do something for me—but the same principle can apply in situations of indirect reciprocity. I do something for you and somebody else helps me in return.

Balzac wrote that behind every large fortune there is a crime. This is a very romantic, absurd, and completely outdated idea. In fact, it very often happens that behind a great fortune or a great success is some action that is particularly generous. In one of my research projects I'm collecting such cases. For example, the French branch of the Rothschild family protected the money of their English clients during the Napoleonic Wars. They were under extreme political pressure to give it up, but they kept the interests of their English clients at heart. Afterward, of course, they became extraordinarily rich because everybody knew that they could be trusted.

From Trivers's work, others derived models about indirect reciprocity, but they were the wrong types of models. People had been reading Axelrod, and there were some abortive attempts at modeling indirect reciprocity and explaining it through game theory. Their conclusion was that reciprocity could not work except in groups of two, which have to interact for a long time. One idea was that the principle behind indirect reciprocity is that if I receive something, I'm more likely to give to the next person who comes along. There might be something true about it, but there have been experiments showing that this principle by itself would not suffice with regard to explaining the stability of indirect reciprocity.

Then a famous scientist named Richard Alexander, a professor at the University of Michigan and a curator of the Museum of Zoology there, wrote a book about the Darwinian evolution of morals. In this book, he asked questions like, What is moral? And how do we start to form our ideas about what is good and bad? We look at what people do for society. We are always assessing the reputations of others, and are more likely to give to somebody who has a high reputation, someone who has in her or his past given help to others—not necessarily to me, but to somebody. If I only give to a person with a high reputation, I channel my help to those who have proved their value for cooperation.

Martin picked up on this work and started with a very simple model. It was a kind of numerical score, a label that says how often a person has given in the past. The idea is that my decision of whether to give to that person or not will depend on that score.

Karl Sigmund

I give more freely to persons with a high score. I refuse to help persons with a low score. This extremely simple model worked very well and inspired a lot of economists to do many experiments based on it.

In this type of experiment, a group of ten people don't know each other and remain anonymous. All they know is that each person in this group has a number, one through ten. Occasionally, two people are chosen randomly. One of them is assigned the role of the donor, and the other is assigned the role of the recipient. The donor can give $3 to the recipient at the cost of $1 dollar to himself. If one assumes this person is selfish and rational, she should not give anything and should keep this $1. She is not going to be punished in any way. In many of these experiments, under conditions of complete anonymity, at the beginning people tend to give once or twice. But when they see there is no immediate return, they stop giving.

However, it's different if the donor knows that next to the number of each participant is a sign indicating how often this participant has given in the past. If a recipient has given five times and refused to give only once, for example, then the recipient has a respectably high score, and it turns out that the donors have a tendency to give preferentially to those with a high score. Milinski and others have studied this example, and it has become almost a trade by now. There are dozens of papers on this simple setup showing that indirect reciprocity works under very simple conditions.

At the same time, though, there have been theoreticians who have said that this model cannot work for a very simple reason: If you are discriminating and see that a recipient is a defector who has not given, then you will not give to that person. But at the same time that you punish him by not giving, your own score will be diminished. Even if this act of not giving is fully justified in your eyes, the next person who sees only whether you have given or not in the past will infer, "Aha! You have not given, and therefore you are a bad guy, and you will not get anything from me." By punishing somebody, you lower your own score and therefore you risk not receiving a benefit from a third person in the next round. Punishing somebody is a costly business, because it costs you future benefits.

The theoreticians then asked, Why should you engage in this act of punishment when it costs you something? This has been called a social dilemma. Punishing others is altruistic in the sense that if you didn't have this possibility of punishment, cooperation would vanish from the group. But this altruism costs you something. Your own reasoning should tell you that it's better to always give, because then your score will always be at a maximum. Therefore, your chances of receiving something will also be maximal.

I'm often thinking about the different ways of cooperating, and nowadays I'm mostly thinking about these strange aspects of indirect reciprocity. Right now it turns out that economists are excited about this idea in the context of e-trading and e-commerce. In this case, you also have a lot of anonymous interactions, not between the same two people but within a hugely mixed group where you are unlikely ever to meet the same person again. Here the question of trusting the other, the idea of reputation, is particularly important. Google page rankings, the reputations of eBay buyers and sellers, and the Amazon.com reader reviews are all based on trust, and there is a lot of moral hazard inherent in these interactions.

Before I get into specific examples, let's first talk about what we have been accustomed to in the past. In a marketplace in ancient Egypt or in a medieval town, you saw the same person day after day. Either he sold you something or you sold him something, and you grew old together. This is no longer the case. In Internet commerce, you buy something from somebody you have never seen, and you put your trust in an agency when you might only see its e-mail address. Lots of money is involved in these interactions, and there are lots of possibilities for cheating. You have to be sure that you trust the right e-commerce company, or that an offer on eBay is serious and will not give you worthless junk. You have to trust someone to send them money before you get the goods, and when they arrive there's the question of determining if the goods are worth it.

Of course, in this kind of e-commerce, there is also the possibility of rating your partner and reporting whether you were satisfied with him the last time or not. This builds up the reputation of every agent in this game, creating a modern version of reputation in a society where you are not meeting the same person day after day, but meeting a stranger.

I became aware of the implications of these ideas for Internet commerce only a short time ago, because I myself don't use eBay, or buy books on Amazon. It was my students who told me that reputation is now a very interesting issue in this context. When I tried for the nth time to write an introduction to a working paper on the topic of indirect reciprocity, they asked me, Why are you always looking to hominid evolution, to prehistory, when a similar thing is happening right now on the Internet? There are now at least ten papers by economists in the works on these topics that I could mention here, but all of them are in the preparatory stage and none has been published.

Whether or not the pioneers behind Google are aware of these theories coming out of evolutionary biology would be an interesting question to ask. I have no contact with them. But I claim a subliminal credit for the term "Google."

When the Brin family first came to Vienna from Russia, Sergey Brin was three years old. The family stayed with us at our apartment for some time. His father, a friend and colleague, is the mathematician Michael Brin, who specializes in ergodic theory and dynamical systems, and that was also my field at that time. The first thing we offered the Brins when they entered our home was a *Guglhupf*, the famous Austrian dessert . . . which greatly impressed the young Sergey. But I bet he does not remember. Officially, the name Google comes from googol—a very large number—and the real reason is lost in oblivion.

Of course, in the early 1980s, I was not thinking about the Internet. I was reading Richard Alexander's work about the evolution of morals and how people first started assessing each other. Alexander knew of course that different cultures have very different morals. Different cultures also have very different languages, and nevertheless many linguists now generally assume that there is a universal language instinct. Similarly, it is probable that although different cultures have different morals, they are based on a universal moral instinct in the sense that there is a tendency to assess people all the time, even if this occurs according to different models. Maybe these depend on the culture or the civilization, but basically our tendency for assessing others is hardwired.

Assessment hardwiring is something that could also be implemented in the context of e-commerce, because if you see two people and one of them refuses to give a gift to the other person, what do you infer? If you only see one isolated act, you will probably think that a potential donor who did not give is a bad guy. But if the potential recipient has a bad reputation, then you would probably say it is justified not to give to that person.

Something being studied experimentally right now is whether

people really go into the fine details when they observe an interaction. Do they observe it in isolation, and do they really care about the standing of the persons involved? Of course, it would be more sophisticated for an assessment to consider both the moral standing of the donor and the moral standing of the recipient before this interaction took place.

In e-commerce, you would want to quantify this moral standing with only minimal parameters. This would ideally rely on a binary label that is either 0 or 1, because this is easiest to implement. But if one starts thinking about different ways to assess an action between two people, there are many possible combinations of what you consider good and bad. It could be that you consider it good to refuse to help a bad guy, or that you consider it bad to help a bad guy. If you analyze all the possibilities, you come to an amazingly high number of possible morals. When we calculated them, we found that there are actually 4,096 possibilities! Which one of them is really working at a given time? This is a question for experimentation, and experiments in this area are in full swing. I know several groups who are doing this. While I am not an experimentalist, I am becoming increasingly interested in that area.

It would be interesting to ask whether or not the pioneers behind Google, eBay, and Amazon are aware of these theories coming out of evolutionary biology. Reputation is something that is profoundly embedded in our mentality, and we—not just old professors, but everyone—care enormously about it. I have read that the moment when people really get desperate and start running amok is when they feel that they are considered completely worthless in their society.

I should stress that we have been talking here essentially about human nature. The more or less official idea that human beings are selfish and rational—an idea that nobody except economists really took seriously, and now even economists say that they never did—has been totally discredited. There are many experiments that show that spontaneous impulses like the tendency for fairness or acts of sympathy or generosity play a huge role in human life.

Karl Sigmund

9.

Digital Maoism

The Hazards of the New Online Collectivism

Jaron Lanier

Computer scientist; musician; author, You Are Not a Gadget

My Wikipedia entry identifies me (at least this week) as a film director. It is true I made one experimental short film about a decade and a half ago. The concept was awful: I tried to imagine what Maya Deren would have done with morphing. It was shown once at a film festival and was never distributed and I would be most comfortable if no one ever sees it again. In the real world it is easy to not direct films. I have attempted to retire from directing films in the alternative universe that is the Wikipedia a number of times, but somebody always overrules me. Every time my Wikipedia entry is corrected, within a day I'm turned into a film director again. I can think of no more suitable punishment than making these determined Wikipedia goblins actually watch my one small old movie.

Twice in the past several weeks, reporters have asked me about my filmmaking career. The fantasies of the goblins have entered that portion of the world that is attempting to remain real. I know I've gotten off easy. The errors in my Wikipedia bio have been (at least prior to the publication of this article) charming and even flattering.

Reading a Wikipedia entry is like reading the Bible closely. There are faint traces of the voices of various anonymous authors and editors, though it is impossible to be sure. In my particular case, it appears that the goblins are probably members or descendants of the rather sweet old *Mondo 2000* culture linking psychedelic experimentation with computers. They seem to place great importance on relating my ideas to those of the psychedelic luminaries of old (and in ways that I happen to find sloppy and incorrect). Edits deviating from this set of odd ideas that are important to this one particular small subculture are immediately removed. This makes sense. Who else would volunteer to pay that much attention and do all that work?

The problem I am concerned with here is not Wikipedia in itself. It's been criticized quite a lot, especially in the past year, but Wikipedia is just one experiment that still has room to change and grow. At the very least it's a success at revealing what the online people with the most determination and time on their hands are thinking, and that's actually interesting information.

No, the problem is in the way Wikipedia has come to be regarded and used; how it's been elevated to such importance so quickly. And that is part of the larger pattern of the appeal of a new online collectivism that is nothing less than a resurgence of the idea that the collective is all-wise, that it is desirable to have influence concentrated in a bottleneck that can channel the collective with the most verity and force. This is different from representative democracy, or meritocracy. This idea has had dreadful consequences when thrust upon us from the extreme Right or the extreme Left in various historical periods. The fact that it's being reintroduced today by prominent technologists and futurists, people whom in many cases I know and like, doesn't make it any less dangerous.

There was a well-publicized study in *Nature* comparing the ac-

Jaron Lanier

curacy of Wikipedia to *Encyclopædia Britannica*. The results were a toss-up. While there is a lingering debate about the validity of the study, the items selected for the comparison were just the sort that Wikipedia would do well on: science topics that the collective at large doesn't care much about. "Kinetic isotope effect" or "Vesalius, Andreas" are examples of topics that make the *Britannica* hard to maintain, because it takes work to find the right authors to research and review a multitude of diverse topics. But they are perfect for Wikipedia. There is little controversy around these items, plus the Net provides ready access to a reasonably small number of competent specialist graduate-student types possessing the manic motivation of youth.

A core belief of the wiki world is that whatever problems exist in the wiki will be incrementally corrected as the process unfolds. This is analogous to the claims of hyper-libertarians who put infinite faith in a free market, or the hyper-lefties who are somehow able to sit through consensus decision-making processes. In all these cases, it seems to me that empirical evidence has yielded mixed results. Sometimes loosely structured collective activities yield continuous improvements and sometimes they don't. Often we don't live long enough to find out. Later in this essay, I'll point out what constraints make a collective smart. But first, it's important not to lose sight of values just because the question of whether a collective can be smart is so fascinating. Accuracy in a text is not enough. A desirable text is more than a collection of accurate references. It is also an expression of personality.

For instance, most of the technical or scientific information that is in Wikipedia was already on the Web before Wikipedia was started. You could always use Google or other search services to find information about items that are now wikified. In some cases, I have noticed, specific texts get cloned from original sites at universities or labs onto wiki pages. And when that happens, each text loses part of its value. Since search engines are now more likely to point you to the wikified versions, the Web has lost some of its flavor in casual use.

When you see the context in which something was written and you know who the author was beyond just a name, you learn so much more than when you find the same text placed in the anonymous, faux-authoritative, anti-contextual brew of Wikipedia. The question isn't just one of authentication and accountability, though those are important, but something more subtle. A voice should be sensed as a whole. You have to have a chance to sense personality in order for language to have its full meaning. Personal Web pages do that, as do journals and books. Even *Britannica* has an editorial voice, which some people have criticized as being vaguely too "Dead White Men."

If an ironic website devoted to destroying cinema claimed that I was a filmmaker, it would suddenly make sense. That would be an authentic piece of text. But placed out of context in Wikipedia, it becomes drivel.

Myspace is another recent experiment that has become even more influential than Wikipedia. Like Wikipedia, it adds just a little to the powers already present on the Web in order to inspire a dramatic shift in use. Myspace is all about authorship, but it doesn't pretend to be all-wise. You can always tell at least a little about the character of the person who made a Myspace page. But it is very rare indeed that a Myspace page inspires even the slightest confidence that the author is a trustworthy authority. Hurray for Myspace on that count!

Myspace is a richer, multilayered source of information than Wikipedia, although the topics the two services cover barely overlap. If you want to research a TV show in terms of what people think of it, Myspace will reveal more to you than the analogous and enormous entries in Wikipedia.

Wikipedia is far from being the only online fetish site for foolish collectivism. There's a frantic race taking place online to become the most "meta" site, to be the highest-level aggregator, subsuming the identity of all other sites.

The race began innocently enough with the notion of creating directories of online destinations, such as the early incarnations of Yahoo! Then came AltaVista, where one could search using an inverted database of the content of the whole Web. Then came Google, which added page-rank algorithms. Then came the blogs, which varied greatly in terms of quality and importance. This led to meta-blogs such as *Boing Boing*, run by identified humans, which served to aggregate blogs. In all of these formulations, real people were still in charge. An individual or individuals were presenting a personality and taking responsibility.

These Web-based designs assumed that value would flow from people. It was still clear, in all such designs, that the Web was made of people, and that ultimately value always came from connecting with real humans.

Even Google by itself (as it stands today) isn't meta enough to be a problem. One layer of page ranking is hardly a threat to authorship, but an accumulation of many layers can create a meaningless murk, and that is another matter.

In the last year or two the trend has been to remove the scent of people, so as to come as close as possible to simulating the appearance of content emerging out of the Web as if it were speaking to us as a supernatural oracle. This is where the use of the Internet crosses the line into delusion.

Kevin Kelly, the former editor of *Whole Earth Review* and the founding executive editor of *Wired*, is a friend and someone who has been thinking about what he and others call the "hive mind." He runs a website called *Cool Tools* that's a cross between a blog and the old *Whole Earth Catalog*. On *Cool Tools*, the contributors, including me, are not a hive because we are identified.

In March, Kelly reviewed a variety of "consensus Web filters" such as Digg and Reddit that assemble material every day from all the myriad other aggregating sites. Such sites intend to be more meta than the sites they aggregate. There is no person taking responsibility for what appears on them, only an algorithm. The hope seems to be that the most meta site will become the mother of all bottlenecks and receive infinite funding.

That new magnitude of meta-ness lasted only a month. In April, Kelly reviewed a site called popurls.com that aggregates consensus Web-filtering sites . . . and there was a new "most meta." We now are reading what a collectivity algorithm derives from what other collectivity algorithms derived from what collectives chose from what a population of mostly amateur writers wrote anonymously.

Is popurls any good? I am writing this on May 27, 2006. In the past few days, an experimental approach to diabetes management has been announced that might prevent nerve damage. That's huge news for tens of millions of Americans. It is not mentioned on popurls. Popurls does clue us in to this news: "Student sets simultaneous world ice cream-eating record, worst ever ice cream headache." Mainstream news sources all lead today with a serious earthquake in Java. Popurls includes a few mentions of the event, but they are buried within the aggregation of aggregate news sites like Google News. The reason the quake appears on popurls at all can be discovered only if you dig through all the aggregating layers to find the original sources, which are those rare entries actually created by professional writers and editors who sign their names. But at the layer of popurls, the ice cream story and the

Javanese earthquake are at best equals, without context or authorship.

Kevin Kelly says of the popurls site, "There's no better way to watch the hive mind." But the hive mind is for the most part stupid and boring. Why pay attention to it?

Readers of my previous rants will notice a parallel between my discomfort with so-called artificial intelligence and the race to erase personality and be most meta. In each case, there's a presumption that something like a distinct kin to individual human intelligence is either about to appear any minute, or has already appeared. The problem with that presumption is that people are all too willing to lower standards in order to make the purported newcomer appear smart. Just as people are willing to bend over backward and make themselves stupid in order to make an AI interface appear smart (as happens when someone can interact with the notorious Microsoft paper clip), so are they willing to become uncritical and dim in order to make meta-aggregator sites appear to be coherent.

There is a pedagogical connection between the culture of AI and the strange allure of anonymous collectivism online. Google's vast servers and Wikipedia are both mentioned frequently as being the start-up memory for artificial intelligences to come. Larry Page is quoted via a link presented to me by popurls this morning (who knows if it's accurate) as speculating that an AI might appear within Google within a few years. George Dyson has wondered if such an entity already exists on the Net, perhaps perched within Google. My point here is not to argue about the existence of metaphysical entities, but just to emphasize how premature and dangerous it is to lower the expectations we hold for individual human intellects.

The beauty of the Internet is that it connects people. The value is in the other people. If we start to believe the Internet itself is an entity that has something to say, we're devaluing those people and making ourselves into idiots.

Compounding the problem is that new business models for people who think and write have not appeared as quickly as we all hoped. Newspapers, for instance, are on the whole facing a grim decline as the Internet takes over the feeding of the curious eyes that hover over morning coffee and, even worse, classified ads. In the new environment, Google News is for the moment better funded and enjoys a more secure future than most of the rather small number of fine reporters around the world who ultimately create most of its content. The aggregator is richer than the aggregated.

The question of new business models for content creators on the Internet is a profound and difficult topic in itself, but it must at least be pointed out that writing professionally and well takes time and that most authors need to be paid to take that time. In this regard, blogging is not writing. For example, it's easy to be loved as a blogger. All you have to do is play to the crowd. Or you can flame the crowd to get attention. Nothing is wrong with either of those activities. What I think of as real writing, however, writing meant to last, is something else. It involves articulating a perspective that is not just reactive to yesterday's moves in a conversation.

The artificial elevation of all things meta is not confined to online culture. It is having a profound influence on how decisions are made in America.

What we are witnessing today is the alarming rise of the fallacy of the infallible collective. Numerous elite organizations have been swept off their feet by the idea. They are inspired by the rise of Wikipedia, by the wealth of Google, and by the rush of entrepreneurs to be the most meta. Government agencies, top corporate planning departments, and major universities have all gotten the bug.

As a consultant, I used to be asked to test an idea or propose a new one to solve a problem. In the past couple of years, I've often been asked to work quite differently. You might find me and the other consultants filling out survey forms or tweaking edits to a collective essay. I'm saying and doing much less than I used to, even though I'm still being paid the same amount. Maybe I shouldn't complain, but the actions of big institutions do matter, and it's time to speak out against the collectivity fad that is upon us.

It's not hard to see why the fallacy of collectivism has become so popular in big organizations: If the principle is correct, then individuals should not be required to take on risks or responsibilities. We live in times of tremendous uncertainties coupled with infinite liability phobia, and we must function within institutions that are loyal to no executive, much less to any lower level member. Every individual who is afraid to say the wrong thing within his or her organization is safer when hiding behind a wiki or some other meta-aggregation ritual.

I've participated in a number of elite, well-paid wikis and meta-surveys lately and have had a chance to observe the results. I have even been part of a wiki about wikis. What I've seen is a loss of insight and subtlety, a disregard for the nuances of considered opinions, and an increased tendency to enshrine the official or normative beliefs of an organization. Why isn't everyone screaming about the recent epidemic of inappropriate uses of the collective? It seems to me the reason is that bad old ideas look confusingly fresh when they are packaged as technology.

The collective rises around us in multifarious ways. What afflicts big institutions also afflicts pop culture. For instance, it has become notoriously difficult to introduce a new pop star in the music business. Even the most successful entrants have hardly ever made it past the first album in the past decade or so. The exception is *American Idol*. As with Wikipedia, there's nothing wrong with it. The problem is its centrality.

More people appear to vote in this pop competition than in presidential elections, and one reason why is the instant convenience of information technology. The collective can vote by phone or by texting, and some vote more than once. The collective is flattered and it responds. The winners are likable, almost by definition.

But John Lennon wouldn't have won. He wouldn't have made it to the finals. Or if he had, he would have ended up a different sort of person and artist. The same could be said about Jimi Hendrix, Elvis, Joni Mitchell, Duke Ellington, David Byrne, Grandmaster Flash, Bob Dylan (please!), and almost anyone else who has been vastly influential in creating pop music.

As below, so above. The *New York Times*, of all places, has recently published op-ed pieces supporting the pseudo-idea of intelligent design. This is astonishing. The *Times* has become the paper of averaging opinions. Something is lost when *American Idol* becomes a leader instead of a follower of pop music. But when intelligent design shares the stage with real science in the paper of record, everything is lost.

How could the *Times* have fallen so far? I don't know, but I would imagine the process was similar to what I've seen in the consulting world of late. It's safer to be the aggregator of the collective. You get to include all sorts of material without committing to anything. You can be superficially interesting without having to worry about the possibility of being wrong.

Except when intelligent thought really matters. In that case the average idea can be quite wrong, and only the best ideas have lasting value. Science is like that.

The collective isn't always stupid. In some special cases the collective can be brilliant. For instance, there's a demonstrative ritual often presented to incoming students at business schools. In one version of the ritual, a large jar of jellybeans is placed in the front of a classroom. Each student guesses how many beans there are. While the guesses vary widely, the average is usually accurate to an uncanny degree.

This is an example of the special kind of intelligence offered by a collective. It is that peculiar trait that has been celebrated as the "wisdom of crowds," though I think the word "wisdom" is misleading. It is part of what makes Adam Smith's Invisible Hand clever, and is connected to the reasons Google's page-rank algorithms work. It was long ago adapted to futurism, where it was known as the Delphi technique. The phenomenon is real, and immensely useful.

But it is not infinitely useful. The collective can be stupid, too. Witness tulip crazes and stock bubbles. Hysteria over fictitious satanic cult child abductions. Y2K mania. The reason the collective can be valuable is precisely that its peaks of intelligence and stupidity are not the same as the ones usually displayed by individuals. Both kinds of intelligence are essential.

What makes a market work, for instance, is the marriage of collective and individual intelligence. A marketplace can't exist only on the basis of having prices determined by competition. It also needs entrepreneurs to come up with the products that are competing in the first place.

In other words, clever individuals, the heroes of the marketplace, ask the questions that are answered by collective behavior. They put the jellybeans in the jar.

There are certain types of answers that ought not to be provided by an individual. When a government bureaucrat sets a price, for instance, the result is often inferior to the answer that would come from a reasonably informed collective that is reasonably free of manipulation or runaway internal resonances. But when a collective designs a product, you get design by committee, which is a derogatory expression for a reason.

Here I must take a moment to comment on Linux and similar efforts. The various formulations of open or free software are different from Wikipedia and the race to be most meta in important ways. Linux programmers are not anonymous, and in fact personal glory is part of the motivational engine that keeps such enterprises in motion. But there are similarities, and the lack of a coherent voice or design sensibility in an aesthetic sense is one negative quality of both open-source software and Wikipedia.

These movements are at their most efficient while building hidden information-plumbing layers, such as Web servers. They are hopeless when it comes to producing fine user interfaces or user experiences. If the code that ran Wikipedia's user interface were as open as the contents of the entries, it would churn itself into impenetrable muck almost immediately. The collective is good at solving problems which demand results that can be evaluated by uncontroversial performance parameters, but bad when taste and judgment matter.

Collectives can be just as stupid as any individual, and in important cases, stupider. The interesting question is whether it's possible to map out where the one is smarter than the many.

There is a lot of history to this topic, and varied disciplines have lots to say. Here is a quick pass at where I think the boundary between effective collective thought and nonsense lies: The collective is more likely to be smart when it isn't defining its own questions, when the goodness of an answer can be evaluated by a simple result (such as a single numeric value), and when the information system which informs the collective is filtered by a quality control mechanism that relies on individuals to a high degree. Under those circumstances, a collective can be smarter than a person. Break any one of those conditions and the collective becomes unreliable or worse.

Meanwhile, individuals best achieve optimal stupidity on those rare occasions when they are both given substantial powers and insulated from the results of their actions.

If the above criteria have any merit, then there is an unfortunate convergence. The setup for the most stupid collective is also the setup for the most stupid individuals.

Every authentic example of collective intelligence that I am aware of also shows how that collective was guided or inspired by well-meaning individuals. These people focused the collective and in some cases also corrected for some of the common hive-mind failure modes. The balancing of influence between people and collectives is the heart of the design of democracies, scientific communities, and many other long-standing projects. There's a lot of experience out there to work with. A few of these old ideas provide interesting new ways to approach the question of how to best use the hive mind.

The pre-Internet world provides some great examples of how personality-based quality control can improve collective intelligence. For instance, an independent press provides tasty news about politicians by reporters with strong voices and reputations, like the Watergate reporting of Woodward and Bernstein. Other writers provide product reviews, such as Walt Mossberg in the *Wall Street Journal* and David Pogue in the *New York Times*. Such journalists inform the collective's determination of election results and pricing. Without an independent press, composed of heroic voices, the collective becomes stupid and unreliable, as has been demonstrated in many historical instances. (Recent events in America have reflected the weakening of the press, in my opinion.)

Scientific communities likewise achieve quality through a co-operative process that includes checks and balances, and ultimately rests on a foundation of goodwill and "blind" elitism—blind in the sense that ideally anyone can gain entry, but only on the basis of a meritocracy. The tenure system and many other aspects of the academy are designed to support the idea that individual scholars matter, not just the process or the collective.

Another example: Entrepreneurs aren't the only "heroes" of a marketplace. The role of a central bank in an economy is not the same as that of a communist party official in a centrally planned economy. Even though setting an interest rate sounds like the an-swering of a question, it is really more like the asking of a ques-tion. The Fed asks the market to answer the question of how to best optimize for lowering inflation, for instance. While that might not be the question everyone would want to have asked, it is at least coherent.

Yes, there have been plenty of scandals in government, the academy, and the press. No mechanism is perfect, but still, here we are, having benefited from all of these institutions. There cer-tainly have been plenty of bad reporters, self-deluded academic scientists, incompetent bureaucrats, and so on. Can the hive mind help keep them in check? The answer provided by experiments in the pre-Internet world is yes, but only provided some signal pro-cessing is placed in the loop.

Some of the regulating mechanisms for collectives that have been most successful in the pre-Internet world can be understood in part as modulating the time domain. For instance, what if a collective moves too readily and quickly, jittering instead of settling down to provide a single answer? This happens on the most active Wikipedia entries, for example, and has also been seen in some speculation frenzies in open markets.

One service performed by representative democracy is low-pass filtering. Imagine the jittery shifts that would take place if a wiki were put in charge of writing laws. It's a terrifying thing to consider. Super-energized people would be struggling to shift the wording of the tax code on a frantic, never-ending basis. The Internet would be swamped.

Such chaos can be avoided in the same way it already is, albeit imperfectly, by the slower processes of elections and court proceedings. The calming effect of orderly democracy achieves more than just the smoothing out of peripatetic struggles for consensus. It also reduces the potential for the collective to suddenly jump into an overexcited state when too many rapid changes to answers coincide in such a way that they don't cancel each other out. (Technical readers will recognize familiar principles in signal processing.)

Wikipedia has recently slapped a crude low-pass filter on the jitteriest entries, such as "President George W. Bush." There's now a limit to how often a particular person can remove someone else's text fragments. I suspect that this will eventually have to evolve into an approximate mirror of democracy as it was before the Internet arrived.

The reverse problem can also appear. The hive mind can be on the right track, but moving too slowly. Sometimes collectives would yield brilliant results given enough time but there isn't enough time. A problem like global warming would automatically be addressed eventually if the market had enough time to respond to it, for instance. Insurance rates would climb, and so on. Alas, in this case, there isn't enough time, because the market conversation is slowed down by the legacy effect of existing investments. Therefore some other process has to intervene, such as politics invoked by individuals.

Another example of the slow-hive problem: There was a lot of technology developed slowly in the millennia before there was a clear idea of how to be empirical, how to have a peer-reviewed technical literature and an education based on it, and before there was an efficient market to determine the value of inventions. What is crucial to notice about modernity is that structure and constraints were part of what sped up the process of technological development, not just pure openness and concessions to the collective.

Let's suppose that Wikipedia will indeed become better in some ways, as is claimed by the faithful, over a period of time. We might still need something better sooner.

Some wikitopians explicitly hope to see education subsumed by wikis. It is at least possible that in the fairly near future enough communication and education will take place through anonymous Internet aggregation that we could become vulnerable to a sudden dangerous empowering of the hive mind. History has shown us again and again that a hive mind is a cruel idiot when it runs on autopilot. Nasty hive-mind outbursts have been flavored Maoist, fascist, and religious, and these are only a small sampling. I don't see why there couldn't be future social disasters that appear suddenly under the cover of technological utopianism. If wikis are to gain any more influence they ought to be improved by mechanisms like the ones that have worked tolerably well in the pre-Internet world.

The hive mind should be thought of as a tool. Empowering the collective does not empower individuals—just the reverse is true. There can be useful feedback loops set up between individuals and the hive mind, but the hive mind is too chaotic to be fed back into itself.

These are just a few ideas about how to train a potentially dangerous collective and not let it get out of the yard. When there's a problem, you want it to bark but not bite you.

The illusion that what we already have is close to good enough, or that it is alive and will fix itself, is the most dangerous illusion of all. By avoiding that nonsense, it ought to be possible to find a humanistic and practical way to maximize the value of the collective on the Web without turning ourselves into idiots. The best guiding principle is to always cherish individuals first.

10.

On Jaron Lanier's "Digital Maoism"

An *Edge* Conversation

With Douglas Rushkoff, Yochai Benkler, Clay Shirky, Cory Doctorow, Kevin Kelly, Esther Dyson, Larry Sanger, Jimmy Wales, and George Dyson

Introduction by Clay Shirky

When Jaron Lanier's piece "Digital Maoism" [see chapter 9] first went out on *Edge*, I knew he'd be generating hundreds of responses all over the Net. After talking to John Brockman, we decided to try to capture some of the best responses here.

Lanier's piece hits a nerve because human life always exists in tension between our individual and group identities, inseparable and incommensurable. For ten years now, it's been apparent that the rise of the digital was providing enormous new powers for the individual. It's now apparent that the world's networks are providing enormous new opportunities for group action.

Understanding how these cohabiting and competing revolutions connect to deep patterns of intellectual and social work is one of the great challenges of our age. The breadth and depth of the responses collected here, ranging from broad philosophical questions to reckonings of the ground truth of particular technologies, is a testament to the complexity and subtlety of that challenge.

Douglas Rushkoff

Media analyst; documentary writer; author, Program, or Be Programmed

Despite comparing Wikipedia with the likes of *American Idol*, this is a more reasoned and hopeful argument than it appears at first glance. Lanier is not condemning collective, bottom-up activity so much as trying to find ways to check its development. In short, it's an argument for the mindful intervention of individuals in the growth and acceleration of this hive-mind thing called collective intelligence.

Indeed, having faith in the beneficence of the collective is as unpredictable as having blind faith in God or a dictator. A poorly developed group mind might well decide any one of us is a threat to the mother organism deserving of immediate expulsion.

Still, I have a hard time fearing that the participants of Wikipedia or even the call-in voters of *American Idol* will be in a position to remake the social order anytime soon. And I'm concerned that any argument against collaborative activity look fairly at the real reasons why some efforts turn out the way they do. Our fledgling collective intelligences are not emerging in a vacuum, but on media platforms with very specific biases.

First off, we can't go on pretending that even our favorite disintermediation efforts are revolutions in any real sense of the word. Projects like Wikipedia do not overthrow any elite at all, but merely replace one elite—in this case an academic one—with another: the interactive-media elite. Just because the latter might include a fourteen-year-old with an Internet connection in no way changes the fact that he's educated, techno-savvy, and enjoying enough free time to research and post to an encyclopedia for no pay. Although he is not on the editorial board of the *Encyclopædia Britannica*, he's certainly in as good a position as anyone to get there.

While I agree with Lanier and the recent spate of articles questioning the confidence so many Internet users now place in user-created databases, these are not grounds to condemn bottom-up networking as a dangerous and headless activity—one to be equated with the doomed mass actions of former communist regimes.

Kevin Kelly's overburdened "hive mind" metaphor notwithstanding, a networked collaboration is not an absolutely level playing field inhabited by drones. It is an ecology of interdependencies. Take a look at any of these online functioning collective intelligences—from eBay to Slashdot—and you'll soon get a sense of who has gained status and influence. And in most cases, these reputations have been won through a process much closer to meritocracy, and through a fairer set of filters, than the ones through which we earn our graduate degrees.

While it may be true that a large number of current websites and group projects contain more content aggregation (links) than original works (stuff), that may as well be a critique of the entirety of Western culture since postmodernism. I'm as tired as anyone of art and thought that exists entirely in the realm of context and reference—but you can't blame Wikipedia for architecture based on winks to earlier eras or a music culture obsessed with sampling old recordings instead of playing new compositions.

Honestly, the loudest outcry over our Internet culture's inclination toward reframing and the "meta" tend to come from those with the most to lose in a society where "credit" is no longer a paramount concern. Most of us who work in or around science and technology understand that our greatest achievements are not personal accomplishments but lucky articulations of collective realizations. Something in the air. (Though attributed to just two men, discovery of the DNA double helix was the result of many groups working in parallel, and no less a collective effort than the Manhattan Project.) Claiming authorship is really just a matter of ego and royalties. Even so, the collective is nowhere near being able to compose a symphony or write a novel—media whose very purpose is to explode the boundaries between the individual creator and his audience.

If you really want to get to the heart of why groups of people using a certain medium tend to behave in a certain way, you'd have to start with an exploration of *biases* of the medium itself. Kids with computers sample and recombine music because computers are particularly good at that—while not so very good as performance instruments. Likewise, the Web—which itself was created to foster the linking of science papers to their footnotes—is a platform biased toward drawing connections between things, not creating them. We don't blame the toaster for its inability to churn butter.

That's why it would be particularly sad to dismiss the possibilities for an emergent collective intelligence based solely on the early results of one interface (the Web) on one network (the Internet) of one device (the computer). The hive-mind metaphor was just one early, optimistic futurist's way of explaining a kind of behavior he hadn't experienced before: that of a virtual community.

Now sure, there may have been a few too many psychedelics making their way through Silicon Valley at the same time as Mac Classics and copies of James Gleick's *Chaos*. At the early breathless phase of any cultural renaissance, there are bound to be some tel-

Douglas Rushkoff

eologically suspect prognostications from those who are pioneering the fringe. And that includes you and me both.

Still, what you saw so clearly from the beginning is that the
beauty of the Internet is its ability to connect people to one another. It's not the content, it's the *contact*.

The Internet itself holds no philosopher's stone—there's no
God to emerge from the medium. I'm with Lanier there. But there
is something that can emerge from people engaging with one another in ways they hadn't dreamed possible before. While the Internet itself may never produce the genuinely cooperative society
so many of us yearn for, it does give us the opportunity to *model*
the kinds of behaviors that may work back here in the real world.
In any case, the true value of the collective is not its ability to
go meta or to generate averages but rather, quite the opposite, to
connect strangers. Already, new subclassifications of diseases have
been identified when enough people with seemingly unique symp-
toms have found one another online. Craigslist's founder is a hero
online not because he has gone meta but because of the very real
and practical connections he has fostered between people looking
for jobs, homes, or families to adopt their pets. And it wasn't Craig
Newmark's intellectual framing that won him this reputation, but
the time and energy he put into maintaining the *social* cohesion of
his online space.

Meanwhile, offline collectivist efforts at disintermediating formerly top-down systems are also creating new possibilities for
everything from economics to education. Local currencies give
unemployed Japanese people the opportunity to spend time caring
for elders near their homes so that someone else can care for their
own family members in distant regions. The New York public
school system owes any hope of a future to the direct intervention

of community members, whose commune-era utopian "free school" models might make us hardened cynics cringe—but energize teachers and students alike.

I'm troubled by *American Idol* and the increasingly pandering *New York Times* as much as anyone, but I don't blame collaboration or techno-utopianism for their ills. In these cases, we're not watching the rise of some new dangerous form of digital populism, but the replacement of key components of a cultural ecology—music and journalism—by the priorities of consumer capitalism.

In fact, the alienating effects of mass marketing are in large part what motivate today's urge toward collective activity. If anything, the rise of online collective activity is itself a check—a low-pass filter on the anti-communal effects of political corruption, market forces, and strident individualism.

One person's check is another person's balance.

The "individual" Lanier would have govern the collective is itself a social construction born in the Renaissance, celebrated via democracy in the Enlightenment, and since devolved into the competition, consumption, and consumerism we endure today.

While the tags adorning Flickr photographs may never constitute an independently functioning intelligence, they do allow people to participate in something bigger than themselves, and they foster a greater understanding of the benefits of collective action. They are a desocialized society's first baby steps toward acting together with more intelligence than people can alone.

And watching for signs of such intelligent life is anything but boring.

Douglas Rushkoff

Yochai Benkler

Berkman Professor of Entrepreneurial Legal Studies, Harvard University; author, The Wealth of Networks: How Social Production Transforms Markets and Freedom

I agree with much of what Jaron Lanier has to say in his insightful essay. The flashy title and the conflation of arguments, however, conspire to suggest that he offers a more general attack on distributed, cooperative networked information production, or what I have called peer production, than Lanier in fact offers.

What are the points of agreement? First, Lanier acknowledges that decentralized production can be effective at certain tasks. In these he includes science-oriented definitions in Wikipedia, where the platform more easily collates the talents, availability, and diverse motivations throughout the network than a slower-moving organization like *Britannica* can; and free and open-source software, though perhaps more in some tasks that are more modular and require less of an overall unifying aesthetic, such as interface. Second, he says these do not amount to a general "collective is always better," but rather to a system that itself needs to be designed to guard against mediocre or malicious contributions through implementation of technical fixes, what he calls "low-pass filters." These parallel the central problem characterized by the social software design movement, as one can see in Clay Shirky's work. Those familiar with my own work in "Coase's Penguin" and since will notice that I only slightly modified Lanier's language to show the convergence of claims. Where, then, is the disagreement?

Lanier has two driving concerns. The first is deep: loss of individuality, devaluation of the unique, responsible, engaged individual as the core element of a system of information, knowledge, and culture. The second strikes me as more superficial, or at least as more time- and space-bound: the concern with the rise of constructs like "hive minds" and metafilters and efforts to build business models around them.

Like Lanier, I see individuals as the bearers of moral claims and the sources of innovation, creativity, and insight. Unlike Lanier, I have argued that enhanced individual practical capabilities represent the critical long-term shift introduced by the networked information economy, improving on the operation of markets and governments in the preceding century and a half. This is where I think we begin to part ways. Lanier has too sanguine a view of markets and governments. To me, markets, governments (democratic or otherwise), social relations, and technical platforms are all various and partly overlapping systems within which individuals exist. They exhibit diverse constraints and affordances, and enable and disable various kinds of action for the individuals who inhabit them. Because of cost constraints and organizational and legal adaptations over the past 150 years, our information, knowledge, and cultural production system has taken on an industrial form, to the exclusion of social and peer production. Britney Spears and *American Idol* are the apotheosis of that industrial information economy, not of the emerging networked information economy.

So too is the decline he decries for the *New York Times*. In my recent work, I have been trying to show how the networked public sphere improves upon the mass mediated public sphere along precisely the dimensions of Fourth Estate function that Lanier extolls, and how the distributed blogosphere can correct, sometimes at least, the mass media failings. It was, after all, Russ Kick's *Memory Hole*, not the *New York Times*, that first broke pictures of military personnel brought home in boxes from Iraq. It was one activist, Bev Harris, with her website Black Box Voting; an academic group led by Avi Rubin; a few Swarthmore students; and a network of thousands who replicated the materials about Diebold voting machines after 2002 that led to review and recall of many voting machines in California and Maryland. The mainstream media, meanwhile, sat by, dutifully repeating the reassurances of officials who bought the machines and vendors who sold them. By now, claims that the Internet democratizes are old.

Going beyond the 1990s' naïve views of democracy in cyberspace, on the one hand, and the persistent fears of fragmentation and the rise of Babel, on the other hand, we can now begin to interpret the increasing amount of data we have on our behavior on the Web and in the blogosphere. What we see in fact is that we are not intellectual lemmings. We do not meander about in the intellectual equivalent of Brownian motion. We cluster around topics we care about. We find people who care about similar issues. We talk. We link. We see what others say and think. And through our choices we develop a different path for determining what issues are relevant and salient, through a distributed system that, while imperfect, is less easily corrupted than the advertising-supported media that dominated the 20th century.

Wikipedia captures the imagination not because it is so perfect, but because it is reasonably good in many cases: a proposition that would have been thought preposterous a mere half decade ago. That it is now compared not to the mainstream commercial encyclopedias like *Grolier's*, *Encarta*, or *Columbia*, but to the quasi-commercial, quasi-professional gold standard of the *Britannica*, is itself the amazing fact. It is, after all, the product of tens of thousands of mostly well-intentioned individuals, some more knowledgeable than others, but almost all flying in the face of Homo economicus and the Leviathan combined. Wikipedia is not faceless, by and large. Its participants develop, mostly, persistent identities (even if not by their real names) and communities around the definitions.

They may not be a perfect, complete replacement for *Britannica*. But they are an alternative, with different motivations, accreditation, and organization. They represent a new solution space to a set of information production problems that we need to experiment with, learn, and develop, but they offer a genuinely alternative form of production than do markets, firms, or governments, and as such an uncorrelated or diverse system of action in the information environment. Improvements in productivity and freedom inhere in this diversity of systems available for human action, not in a generalized claim of superiority for one of these systems over all the others under all conditions.

This leaves the much narrower set of moves that are potentially the legitimate object of Lanier's critique: efforts that try to depersonalize the "wisdom of crowds," unmooring it from the individuals who participate; try to create ever-higher-level aggregation and centralization in order to "capture" that "wisdom"; or imagine it as emergent in the Net, abstracted from human minds. I'm not actually sure there is anyone who genuinely holds such a hyperbolic version of this view. I will, in any event, let others defend it if they do hold such a view.

Here I will only note that the centralized filters Lanier decries are purely an effort to re-create price-like signaling in a context—information in general, and digital networks in particular—where the money-based price system is systematically dysfunctional. It may be right- or wrongheaded, imperfect or perfect. But it is not collectivism.

Take Google's algorithm. It aggregates the distributed judgments of millions of people who have bothered to host a webpage.

It doesn't take just any judgment, only those that people care enough about to exert the effort to insert a link in their own page to some other page. In other words, relatively "scarce" or "expensive" choices. It doesn't ask the individuals to submerge their identity, or preferences, or actions in any collective effort. No one spends their evenings in consensus-building meetings. It merely produces a snapshot of how they spend their scarce resources: time, webpage space, expectations about their readers' attention. That is what any effort to synthesize a market price does. Anyone who claims to have found transcendent wisdom in the pattern emerging from how people spend their scarce resources is a follower of Milton Friedman, not of Chairman Mao.

At that point, Lanier's critique could be about the way in which markets of any form quash individual creativity and unique expression; it might be about how excessive layers of filtering degrade the quality of information extracted from people's behavior with their scarce resources, so that these particular implementations are poor market-replacement devices. In either case, his lot is with those of us who see the emergence of social production and peer production as an alternative to both state-based and market-based, closed proprietary systems, thereby enhancing creativity, productivity, and freedom.

To conclude: The spin of Lanier's piece is wrong. Much of the substance is useful. The big substantive limitation I see is his excessively rosy view of the efficacy of the price system in information production. Networked-based, distributed social production, both individual and cooperative, offers a new system, alongside markets, firms, governments, and traditional nonprofits, within which individuals can engage with information, knowledge, and cultural production. This new modality of production offers new challenges and new opportunities. It is the polar opposite of

Maoism. It is based on enhanced individual capabilities, employing widely distributed computation, communication, and storage in the hands of individuals with insight, motivation, and time, and deployed at their initiative through technical and social networks, either individually or in loose voluntary associations.

Clay Shirky

Social and technology network topology researcher; adjunct professor, New York University Graduate School, Interactive Telecommunications Program (ITP); author, Cognitive Surplus

Jaron Lanier is certainly right to look at the downsides of collective action. It's not a revolution if nobody loses, and in this case, expertise and iconoclasm are both relegated by some forms of group activity. However, "Digital Maoism" mischaracterizes the present situation in two ways. The first is that the target of the piece, the "hive mind," is just a catchphrase used by people who don't understand how things like Wikipedia really work. As a result, criticism of the hive mind becomes similarly vague. Second, the initial premise of the piece—there are downsides to collective production of intellectual work—gets spread so widely that it comes to cover RSS aggregators, *American Idol,* and the editorial judgment of the *New York Times.* These are errors of overgeneralization; it would be good to have a conversation about Wikipedia's methods and governance, say, but that conversation can't happen without talking about its actual workings, nor can it happen if it is casually lumped together with other, dissimilar kinds of group action.

The bigger of those two mistakes appears early: "The problem I am concerned with here is not Wikipedia in itself. It's been criticized quite a lot, especially in the past year, but Wikipedia is just one experiment that still has room to change and grow. . . . No, the problem is in the way Wikipedia has come to be regarded and used; how it's been elevated to such importance so quickly." Curiously, the ability of the real Wikipedia to adapt to new challenges is taken at face value. The criticism is then directed instead at people proclaiming Wikipedia as an avatar of a golden era of collective consciousness. Let us stipulate that people who use terms like "hive mind" to discuss Wikipedia and other social software are credulous at best, and that their pronouncements tend toward caricature. What "Digital Maoism" misses is that Wikipedia doesn't work the way those people say it does.

Neither proponents nor detractors of hive-mind rhetoric have much interesting to say about Wikipedia itself, because both groups ignore the details. As Fernanda Viégas's work shows, Wikipedia isn't an experiment in anonymous collectivist creation; it is a specific form of production, with its own bureaucratic logic and processes for maintaining editorial control. Indeed, though the public discussions of Wikipedia often focus on the "everyone can edit" notion, the truth of the matter is that a small group of participants design and enforce editorial policy through mechanisms like the Talk pages, lock protection, article-inclusion voting, mailing lists, and so on. Furthermore, proposed edits are highly dependent on individual reputation—anonymous additions or alterations are subjected to a higher degree of both scrutiny and control, while the reputation of known contributors is publicly discussed on the Talk pages.

Wikipedia is best viewed as an engaged community that uses a large and growing number of regulatory mechanisms to manage a huge set of proposed edits. "Digital Maoism" specifically rejects that point of view, setting up a false contrast with open-source projects like Linux, when in fact the motivations of Wikipedia contributors are much the same. With both systems, there are a huge number of casual contributors and a small number of dedicated maintainers, and in both systems part of the motivation comes from appreciation of knowledgeable peers rather than the general public. Contra Lanier, individual motivations in Wikipedia are not only alive and well, it would collapse without them.

The "Digital Maoism" argument is further muddied by the other systems dragged in for collectivist criticism. There's the inclusion of *American Idol*, in which a popularity contest is faulted for privileging popularity. Well, yes, it would, wouldn't it, but the negative effects here don't come from some new form of collectivity, they come from voting, a tool of fairly ancient provenance. Decrying *Idol*'s centrality is similarly misdirected. This season's final episode was viewed by roughly a fifth of the country. By way of contrast, the final episode of *M*A*S*H* was watched by *three-fifths* of the country. The centrality of TV, and indeed of any particular medium, has been in decline for three decades. If the pernicious new collectivism is relying on growing media concentration, we're safe.

Popurls.com is similarly and oddly added to the argument, but there is in fact no meta-collectivity algorithm at work there— Popurls is just an aggregation of RSS feeds. You might as well go after My.Yahoo.com if that's the kind of thing that winds you up. And the ranking systems that are aggregated all display different content, suggesting real subtleties in the interplay of algorithm and audience, rather than a homogenizing hive mind at work. You wouldn't know it, though, to read the broad-brush criticism of Popurls here. And that is the missed opportunity of "Digital Maoism": There are things wrong with RSS aggregators, ranking algorithms, group editing tools, and voting, things we should identify and try to fix. But the things wrong with voting aren't wrong with editing tools, and the things wrong with ranking algorithms aren't wrong with aggregators. To take the specific case of Wikipedia, the Seigenthaler/Kennedy debacle catalyzed both soul-searching and new controls to address the problems exposed, and the controls included, *inter alia*, a greater focus on individual responsibility, the very factor "Digital Maoism" denies is at work.

The changes we are discussing here are fundamental. The personal computer produced an incredible increase in the creative autonomy of the individual. The Internet has made group-forming ridiculously easy. Since social life involves a tension between individual freedom and group participation, the changes wrought by computers and networks are therefore in tension. To have a discussion about the pluses and minuses of various forms of group action, though, is going to require discussing the current tools and services as they exist, rather than discussing their caricatures or simply wishing that they would disappear.

Cory Doctorow

Science-fiction novelist; blogger; technology activist; coeditor, Boing Boing; *author,* Makers

Where Jaron Lanier sees centralization, I see decentralization. Wikipedia is notable for lots of reasons, but the most interesting one is that Wikipedia—a genuinely useful information resource of great depth and breadth—was created in almost no time, at almost no cost, by people who had no access to the traditional canon.

We're bad futurists, we humans. We're bad at predicting what will be important and useful tomorrow. We think the telephone will be best used to bring opera to America's living rooms. We set out nobly to make TV into an educational medium. We create functional hypertext to facilitate the sharing of draft physics papers.

If you need to convince a gatekeeper that your contribution is worthy before you're allowed to make it, you'd better hope the gatekeeper has superhuman prescience. (Gatekeepers don't have superhuman prescience.) Historically, the best way to keep the important things rolling off the lines is to reduce the barriers to entry. Important things are a fraction of all things, and therefore, the more things you have, the more important things you'll have.

The worst judges of tomorrow's important things are today's incumbents. If you're about to creatively destroy some incumbent's business model, that incumbent will be able to tell you all kinds of reasons why you should cut it out. Travel agents had lots of soothing platitudes about why Expedia would never fly. Remember travel agents? Wonder how that worked out for them.

The travel agents were right, of course. Trying to change your own plane tickets stinks. But Internet travel succeeds by being good at the stuff that travel agents sucked at, not good at the stuff that made travel agents great. Internet travel is great because it's cheap and always on, because you can reclaim the "agency" (ahem) of plotting your route and seeing the timetables, and because you can comparison-shop in a way that was never possible before.

Wikipedia isn't great because it's like *Britannica*. *Britannica* is great at being authoritative, edited, expensive, and monolithic. Wikipedia is great at being free, brawling, universal, and instantaneous.

Making a million-entry encyclopedia out of photons, philosophy, and peer pressure would have been impossible before the

Internet's "collectivism." Wikipedia is a noble experiment in defining a protocol for organizing the individual efforts of disparate authors with conflicting agendas. Even better, it has a meta-framework—its GNU copyright license—that allows anyone else to take all that stuff and use part or all of Wikipedia to seed different approaches to the problem.

Wikipedia's voice is by no means bland, either. If you suffice yourself with the actual Wikipedia entries, they can be a little papery, sure. But that's like reading a mailing list by examining nothing but the headers. Wikipedia *entries* are nothing but the emergent effect of all the angry thrashing going on below the surface.

No, if you want to really navigate the truth via Wikipedia, you have to dig into those "history" and "discuss" pages hanging off of every entry. That's where the real action is, the tidily organized palimpsest of the flame war that lurks beneath any definition of "truth."

Britannica tells you what dead white men agreed upon; Wikipedia tells you what live Internet users are fighting over.

Britannica truth is an illusion, anyway. There's more than one approach to any issue, and being able to see multiple versions of them, organized with argument and counter-argument, will do a better job of equipping you to figure out which truth suits you best.

True, reading Wikipedia is a media literacy exercise. You need to acquire new skill-sets to parse out the palimpsest. That's what makes it genuinely novel. Reading Wikipedia like *Britannica* stinks. Reading Wikipedia like Wikipedia is mind-opening.

Free software like Ubuntu Linux and Firefox can have beautiful UIs (despite Lanier's claims), and the authors who made those UIs and their codebase surely put in that work for the egoboo and credit. But you'll never know who designed your favorite UI widget unless you learn to read the Firefox palimpsest: the source tree.

Wikipedia doesn't supplant individual voices like those on blogs. Wikipedia contributors are often prolific bloggers, wont to talk about their work on Wikipedia in LiveJournals and TypePads and WordPresses. Wikipedia is additive—it creates an additional resource out of the labor of those passionate users.

So Wikipedia gets it wrong. *Britannica* gets it wrong, too. The important thing about systems isn't how they work; it's how they fail. Fixing a Wikipedia article is simple. Participating in the brawl takes more effort, but then, that's the price you pay for truth, and it's still cheaper than starting up your own *Britannica*.

Kevin Kelly

Editor at large, Wired; *author,* What Technology Wants

Wikipedia is all that it claims to be: a free encyclopedia created by its readers, that is, by anyone on the Internet. That feat would be wonderful enough, but its origin is so peculiar, and its existence so handy, the obvious follow-up question has become, Is it anything else? Is Wikipedia a template for other kinds of information, or maybe even other kinds of creative works? Is the way Wikipedia is authored a guide to the way many new things might be created? Is it something we should aim toward? Is it a proxy of what is coming in the century ahead?

That's a heavy mythic load to put on something only a few years old, but it seems to have stuck. For better or worse, Wikipedia now represents smart chaos, or bottom-up power, or decentralized being, or out-of-control goodness, or what I seem to have called, for the lack of a better term, the hive mind. It is not the only hive mind out there. We see the Web itself, and other collective entities such as fandoms, voting audiences, link aggregators, consensus filters, open-source communities, and so on, all basking in a rising tide of loosely connected communal action.

But it doesn't take very long to discover that none of these innovations is pure hive mind, and that the supposed paragon of ad hocary—Wikipedia—is itself far from strictly bottom-up. In fact a close inspection of Wikipedia's process reveals that it has an elite at its center (that it has a center is news to most) and that there is far more deliberate design management going on than first appears.

This is why Wikipedia has worked in such a short time. The main drawback to pure, unadulterated Darwinism is that it takes place in biological time—eons. The top-down design part woven deep within by Jimmy Wales and associates has allowed Wikipedia to be smarter than pure dumb evolution would allow in a few years. It is important to remember how dumb the bottom is in essence. In biological natural selection, the prime architect is death. What's dumber than that? One binary bit.

We are too much in a hurry to wait around for a pure hive mind. Our technological systems are marked by the fact that we have introduced intelligent design into them. This is the top-down control we insert to speed and direct a system toward our goals. Every technological system, including Wikipedia, has design in it. What's new is only this: Never before have we been able to make systems with as much "hive" in them as we have recently made with the Web. Until this era, technology was primarily all control, all design. Now it can be design and hive. In fact, this Web 2.0 business is chiefly the first step in exploring all the ways in which we can combine design and the hive in innumerable permutations. We are tweaking the dial in hundreds of combos: dumb writers, smart filters; smart writers, dumb filters, ad infinitum.

But if the hive mind is so dumb, why bother with it at all?

Because as dumb as it is, it is smart enough. More importantly, its brute dumbness produces the raw material that design smarts can work on. If we listened *only* to the hive mind, that would be stupid. But if we ignore the hive mind altogether, that is even stupider.

There's a bottom to the bottom. I hope we realize that a massive bottom-up effort will only take us partway—at least in human time. That's why it should be no surprise to anyone that over time, more and more design, more and more control, more and more structure will be layered into Wikipedia. I would guess that in fifty years a significant portion of Wikipedia articles will have controlled edits, peer review, verification locks, authentication certificates, and so on. That's all good for us readers. The fast-moving frontiers will probably be as open and wild as they are now. That's also great for us.

Futhermore, I know it is heresy, but it might be that the Wikipedia model is not good for very much more than writing universal encyclopedias. Perhaps the article length is fortuitously exactly the right length for the smart mob, and maybe a book is exactly the wrong length. However, while the 2006 Wikipedia process may not be the best way to make a textbook, or create the encyclopedia of all species, or dispense the news, the 2056 Wikipedia process, with far more design in it, may be.

It may be equally heretical to suggest that the hive mind will write far more of our textbooks and databases and news than anyone might believe right now.

Here's how I sum it up: The bottom-up hive mind will always

take us much further than seems possible. It keeps surprising us. In this regard, Wikipedia truly is exhibit A, impure as it is, because it is something that is impossible in theory, and only possible in practice. It proves the dumb thing is smarter than we think. At the same time, the bottom-up hive mind will never take us to our end goal. We are too impatient. So we add design and top-down control to get where we want to go.

Judged from where we start, harnessing the dumb power of the hive mind will take us much further than we can dream. Judged from where we end up, the hive mind is not enough; we need top-down design.

Since we are only at the start of the start, it's the hive mind all the way for now.

Long live Wikipedia!

Esther Dyson

Catalyst, information technology start-ups, EDventure Holdings; author, Release 2.1

I'll just be short, since I'm too busy reading hive-mind output: I think the real argument is between voting or aggregating—where anonymous people raise or lower things in esteem by the weight of sheer numbers—vs. arguments by recognizable individuals that answer the arguments of other individuals. . . . The first is useful in coming up with numbers and trends and leading movements, but it's not creative in the way that evolution, for example, creates species. Evolution isn't blind voting. It works by using a grammar (of genetic materials and unfolding proteins; some biologist will correct me here, for sure) to make changes that are consistent with the whole—i.e., adding two new limbs at a time, or adding muscle to support added mass. Arguments may be won or lost and a consensus argument or belief may arise, but it is structured and emerges more finely shaped than what mere voting or "collectivism" would have produced.

That's why we have representative government—in theory at least. Certain people—designated "experts"—sit together to design something that is supposed to be coherent. (That's the vision, anyway.) You can easily vote both for lower taxes and more services, but you can't design a consistent system that will deliver that.

So, to get the best results, we have people sharpening their ideas against one another rather than simply editing someone's contribution and replacing it with another. We also have a world where the contributors have identities (real or fake, but consistent and persistent) and are accountable for their words. Much like *Edge*, in fact.

———

Larry Sanger

Cofounder, Wikipedia; director of collaborative projects, Digital Universe Foundation; director, Text Outline Project

What exactly is Jaron Lanier's thesis? His main theme is that a certain kind of collectivism is in the ascendancy, and that's a terrible thing. He decries the view "that the collective is all-wise, that it is desirable to have influence concentrated in a bottleneck that can channel the collective with the most verity and force."

I find myself agreeing with Lanier: The collectivism he describes *is* a terrible thing, by golly, and far too many people we admire seem to be caught up in it. But in agreeing, I find myself in a couple of paradoxes. First, surely, no one would *admit* to believing that "the collective is all-wise." So hasn't Lanier set up a straw man? Second, I myself am an advocate of what I call "strong collaboration," exemplified by Wikipedia, in which a work is developed not just by multiple authors, but a constantly changing battery of authors, none of whom "own" the work. So am I not *myself* committed, if anyone is, to believing "the collective is all-wise"?

To understand Lanier's thesis, and where I agree with it—and why it isn't a straw man—it helps to consider certain attitudes one pretty commonly finds in the likes of Wikipedia, Slashdot, and the blogosphere generally. Let me describe something close to home. In late 2004, I publicly criticized Wikipedia for failing to respect expertise properly, and a surprisingly large number of people replied that, essentially, Wikipedia's success has shown that "experts" are no longer needed, that a wide-ranging description of everyone's opinions is more valuable than what some narrow-minded "expert" thinks.

Slashdot's post-ranking system is another perfect example. Slashdotters simply would not stand for a system in which some hand-selected group of editors chose or promoted posts; but if the result is decided by an impersonal algorithm, then it's okay. It isn't that the Slashdotters have a rational belief that the cream will rise to the top under the system; people use the system just because it seems *fairer* or *more equal* to them.

It's not *quite* right to say the "collectivists" believe that the collective is all-wise. Rather, they don't really *care* about getting it right as much as they care about equality.

You might notice that Lanier never bothered to *refute*, in his essay, the view that the collective is all-wise. That's because this view is obviously wrong. Truth and high quality in general are obviously not guaranteed by sheer numbers. But then the champions of collective opinion-making and aggregation surely don't think they *are*. So isn't Lanier just knocking down a straw man? I don't think so. As I take it, the substance of Lanier's point is that the aggregate views expressed by the collective are actually more *valuable*, in some sense, than anything produced by people designated as "experts" or "authorities."

Think about that a bit. Ultimately, I think there is a deep epistemological issue at work here. Epistemologists have a term, "positive epistemic status," for the positive features that can attach to beliefs; so truth, knowledge, justification, evidence, and various other terms are all names for various kinds of positive epistemic status.

So I think we are discovering that there is a lively movement afoot that rejects the traditional kinds of positive epistemic status and wants to replace them with, or explain them in terms of, whatever it is that the collective (i.e., a large group of people, of which one is a part) believes or endorses. We can give this view a name, for convenience: epistemic collectivism.

Epistemic collectivism is a real phenomenon; whether they admit it or not, a lot of people *do* place the views of the collective uppermost. People are epistemic collectivists in just the same way, and for just the same reasons, that they are abject conformists. Surely epistemic collectivism has its roots in the easy sophomoric embrace of relativism. If there is no objective truth, as so many of my old college students seemed to believe, then there is no way to make sense of the idea of expertise or of intellectual authority. Without a reality "out there," independent of us, that we can be right or wrong about, there is no way to justify placing some "experts" above the rest of us in terms of the reliability of their claims. If you're an epistemic collectivist, then it's natural to think that the experts can be overruled by the rest of us.

Now to the second paradox I mentioned earlier. How can I agree with Lanier and still promote strong collaboration? How can I reject epistemic collectivism yet say that Wikipedia is a great project, which I do? Well, the problem is that epistemic collectivists like Wikipedia *but for the wrong reasons.* What's great about it is *not* that it produces an averaged view that is somehow *better* than an authoritative statement by people who actually know the subject. *That's just not it at all.* What's great about Wikipedia is the fact that it is a way to organize enormous amounts of labor for a single *intellectual* purpose. The virtue of strong collaboration, as demonstrated by projects like Wikipedia, is that it represents a new kind of "industrial revolution," where what is reorganized is not *techné* but instead mental effort. It's the sheer *efficiency* of strongly collaborative systems that is so great, not their ability to produce The Truth. Just *how* to eke The Truth out of such a strongly collaborative system is an unsolved, and largely unaddressed, problem.

So online collaboration in some people's minds *can* be indistinguishable from a new collectivism, and Lanier is right both to say so and to condemn the fact. But this collectivism is inherent neither in tools, such as wikis, nor in methods, such as collaboration and aggregation.

———

Jimmy Wales

Founder and chair emeritus, Board of Trustees, Wikimedia Foundation (nonprofit corporation that operates Wikipedia); cofounder, Wikia, Inc. (for-profit company)

> A core belief of the wiki world is that whatever problems exist
> in the wiki will be incrementally corrected as the process
> unfolds.

My response is quite simple: This alleged "core belief" is not one that is held by me, nor as far as I know, by any important or prominent Wikipedians. Nor do we have any particular faith in collectives or in collectivism as a mode of writing. Authoring at Wikipedia, as everywhere, is done by individuals exercising the judgment of their own minds.

> The best guiding principle is to always cherish individuals
> first.

Indeed.

George Dyson

Science historian; author, Darwin Among the Machines *and* Project Orion

This delightful and much-needed essay is the product of a brilliant individual mind at work.

However, Lanier's high-level insights are themselves the result of exactly those collective, haphazard, and noisy processes that are under criticism here. Deep within Jaron Lanier's brain, layer upon layer of anonymous neurons have cycled collectively through meta-meta-meta levels of information processing to produce the thinking he presents so coherently in words. Underlying everything from music to vision are social networks where popularity and having the right connections wins. When Lanier was in his infancy, processes similar to PageRank, AdSense, and AdWords, running (and competing) amok among billions of neurons and trillions of synapses, allowed the language, symbols, and meaning embodied in his surrounding human culture to take root. When it comes to natural intelligence, Wikipedia, not *Britannica*, wrote the book.

All intelligence is collective. But, as Lanier points out, that does not mean that all collectives are intelligent.

The important part of his message is a warning to respect, and preserve, our own intelligence. The dangers of relinquishing individual intelligence are real.

Lanier does not want to debate the existence or nonexistence of metaphysical entities. But his argument that online collectivism produces artificial stupidity offers no reassurance to me. Real artificial intelligence (if and when) will be unfathomable to us. At our level, it may appear as dumb as *American Idol*, or as pointless as a nervous twitch that corrects and uncorrects Jaron Lanier's Wikipedia entry in an endless loop.

11.
Social Networks Are Like the Eye

Nicholas A. Christakis

Physician and social scientist, Harvard University; coauthor,
Connected: The Surprising Power of Social Networks
and How They Shape Our Lives

There is a well-known example in evolutionary biology about
whether the eye was designed, or is "just so" because it evolved and
arose for a reason. How could this incredibly complicated thing
come into being? It seems to serve an incredibly complicated purpose, and the eye is often used in debates about evolution precisely
because it is so complex and seems to serve such a specialized and
critical function.

For me, social networks are like the eye. They are incredibly
complex and beautiful, and looking at them begs the question of
why they exist, and why they come to pass. Do we need a kind of
Just-so Story to explain them? Do they just happen to be there,
for no particular reason? Or do they serve some purpose—some
ontological and also pragmatic purpose?

Along with my collaborator James Fowler, I have been wres-
tling with the questions of where social networks come from, what
purpose they serve, what rules they follow, and what they mean for
our lives. The amazing thing about *social* networks, unlike other
networks that are almost as interesting—networks of neurons or
genes or stars or computers or all kinds of other things one can
imagine—is that the nodes of a social network—the entities, the

components—are themselves sentient, acting individuals who can respond to the network and actually form it themselves.

In social networks, there is an interdigitation between the higher-order structure and the lower-order structure, which is remarkable, and which has been animating our research for the past five or ten years. I started by studying very simple dyadic networks. A pair of individuals is the simplest type of network one can imagine. And I became curious about networks and network effects in my capacity as a doctor who takes care of people who are terminally ill.

In addition to my training in social science, I was trained as a hospice doctor. When I was at the University of Chicago (until 2001), I had a very special clinical practice that involved taking care of people in their own homes, and on Sunday afternoons I would take my little black bag to the South Side of Chicago and visit people who were dying. I had a sort of schizophrenic practice. About a third of my patients were very educated people associated with the University of Chicago, and two-thirds were indigent people from the South Side.

I have the very distinct image in my mind of experiences of myself driving to a borderline safe community, parking my car, looking around, walking up the short steps to the door, knocking, and waiting for what often seemed like a very long time for someone to come to the door. And then being led into people's homes, often by the spouse of the person who was dying. There were often other relatives around and my primary focus as a hospice doctor was not just the person who was dying, but also the family members. I became increasingly interested in this.

I began to see in a very real way that the illness of the person dying was affecting the health status of other individuals in the family. And I began to see this as a kind of nonbiological trans-

Nicholas A. Christakis

mission of disease—as if illness or death or health care use in one person could cause illness or death or health care use in other people connected to him. It wasn't an epidemic transmission of a germ; something else was happening. This is a very basic observation about what I now call "interpersonal health effects," but as I began to have more and more clinical experience with such patients, I began to broaden the focus. I became interested not just in dyadic transmission of illness and illness burden, but also hyper-dyadic transmission.

For example, one day I met with a pretty typical scenario: a woman who was dying and her daughter who was caring for her. The mother had been sick for quite a while, and she had dementia. The daughter was exhausted from years of caring for her, and in the course of caring, she became so exhausted that her husband also became sick from his wife's preoccupation with her mother. One day I got a call from the husband's best friend, with the husband's permission, to ask me about him. So here we have the following cascade: parent to daughter, daughter to husband, and husband to friend. That is four people—a cascade of effects through the network. And I became sort of obsessed with the notion that these little dyads of people could agglomerate to form larger structures.

Nowadays, most people have these very distinct visual images of networks, because in the past ten years, they have become almost a part of pop culture. But social networks were studied in this kind of way beginning in the 1950s—actually, there was some work done in the 1930s and even earlier by a sociologist by the name of Georg Simmel—and culminating in the 1970s with seminal work that was done by sociologists at that time (people like Mark Granovetter, Stan Wasserman, Ron Burt, and others). But all these were still very small-scale networks; networks of three people or thirty people—that kind of ballpark. But we are of course connected to each other through vastly larger, more complex, more beautiful networks of people. Networks of thousands of individuals, in fact. These networks are in a way living, breathing entities that reproduce, and that have a kind of memory. Things flow through them and they have a purpose and can achieve different things from what their constituent individuals can. And they are very difficult to understand.

This is how I began to think about social networks about seven years ago. At the time when I was thinking about this, I moved from the University of Chicago to Harvard and was introduced to my colleague James Fowler, another social scientist, who was also beginning to think about different kinds of network problems from the perspective of political science. He was interested in problems of collective action—how groups of people are organized, how the action of one individual can influence the actions of other individuals. He was also interested in basic problems like altruism. Why would I be altruistic toward somebody else? What purpose does altruism serve? In fact, I think that altruism is a key predicate to the formation of social networks because it serves to stabilize social ties. If I were constantly violent toward other people, or never reciprocated anything good, the network would disintegrate, all the ties would be cut. Some level of altruism is required for networks to emerge.

So we can begin to think about combining a broad variety of ideas. Some stretch back to Plato, and thinking about well-ordered societies, the origins of good and evil, how people form collectives, how a state might be organized. In fact, we can begin to revisit ideas engaged by Rousseau and other philosophers on man in a state of nature. How can we transcend anarchy? Anarchy can be conceived of as a kind of social network phenomenon, and society and social order can also be conceived of as a social network phenomenon.

Nicholas A. Christakis

We can start with the tiny case of a man and a woman—a pair of individuals—one of whom is sick and the other of whom cares (partly out of altruistic reasons) for that person. Stepping back to see them not as individuals, but focusing on the tie that connects them as the object of inquiry, we see that they are embedded in larger sets of such networks, which forces us to engage with a set of fundamental social scientific and philosophical problems—in fact moral problems—that people have been concerned with for millennia.

There is another aspect to the intellectual history of the study of networks that is very interesting. In the fifties and in the seventies, several social scientists began to study social networks and struggled with the problem of nodes (people) and the ties or "edges" that connect them. In fact, "edge" is the formal network term for the connection between two people on a network "graph."

They began to struggle with how to understand this phenomenon and developed a variety of ideas and statistical methods for studying social networks. They did not have data on a large scale and they were limited by the computational power available to them at that time, but they made a lot of progress. They invented a lot of techniques and pushed the field about as far as it could go then. After that there was a quiescent period; and the initial heyday of social network studies was back in the seventies.

These methods incidentally were built on some efforts by very well-known Hungarian mathematicians who studied a branch of mathematics known as topology, which itself has an interesting and old history stretching back to Euler. Beginning in the 1990s, there was a kind of resurrection of network science, initially caused by a group of physicists and mathematicians who were actually tackling problems in other domains: for instance, people interested in networks of genes, or cellular networks, or networks of neurons, like my colleague László Barabási. If we have, for example, a simple worm that has two hundred neurons, can we map all of the connections between them and thereby understand how the worm learns, or how it behaves? Can we understand learning and behavior not by studying the neurons, but by studying the *interconnection* between neurons?

A lot of scientists became interested in other kinds of networks and latched on to many of the old sociological ideas. They developed the mathematics and applied them in new ways, tremendously improving the science of networks—people like Barabási and Duncan Watts and Steve Strogatz and Mark Newman. Now all of this methodological apparatus is flowing back to the social sciences, and social scientists are using it to revisit and understand again a topic that has been of great concern to them for some time.

We are thus at a moment where a leap forward in the methodology for the study of social networks has been made, first by building on past work. But second, we are at a moment where—because of modern telecommunications technologies and other innovations—people are leaving digital traces of where they are, whom they are interacting with, and what they are saying or even thinking. All of these types of data can be captured by the deployment of what I call "massive passive" technologies and used to engage social science questions in a way that our predecessors could only dream of. We have vast amounts of data that can be reapplied to investigate fundamental questions about social organization and about morality and other concerns that have perplexed us forever.

We have had advances in methods; we have had advances in data. We have also had advances in ideas. People are beginning to think more creatively about what it means to have these kinds of higher-order structures. Since the late 1990s and on into the

Nicholas A. Christakis

2000s, science more generally has been engaged in what I call the "assembly project" of modern science. Astronomers are beginning to think about how to assemble stars into galaxies, computer scientists are thinking about how to assemble computers into networks. With the rapid development of the Internet in the mid–1990s, everybody began to think about computers and their networks, and about how they interact and so forth. Engineers struggle with these problems.

Neuroscientists are beginning to think, Okay, well, we understand a lot about neurons, but how do they interconnect to form brains? Geneticists are saying, At the end of the day, we will have understood all 25,000 (approximately) human genes, and then what? How do we put Humpty Dumpty back together again? How do we reassemble all of the genes and understand how they interact with each other in space and across time? We have seen the recent birth of a new field of biology called systems biology, which seeks to put the parts back together.

And similarly, in social science, there is an increasing interest in the same kind of phenomenon. We have begun to understand human behavior, and we have models of rational decision-making—rational actor models—that have led to further innovations. But these models all pertain primarily to individuals. Adam Smith talked about markets as a phenomenon that emerges from the action of individuals, but nevertheless we have primarily focused on the actions of individuals. How do we put all these parts back together to understand groups? Again, the study of social networks is part of this assembly project, part of this effort to understand how you can then have the emergence of order and the emergence of new phenomena that do not inhere in the individuals. We have, for example, consciousness, which cannot be understood by studying neurons. Consciousness is an emergent property of neuronal tissue. And we can imagine similarly certain kinds of emergent properties of social networks that do not inhere in the individuals—properties that arise because of the ties between individuals and because of the complexity of those ties.

Understanding all of this is what drives me and James Fowler to death right now. And as we have been thinking about it, we have come up with some initial simple ideas, and some initial intriguing and very novel empirical observations. The simple ideas are the following: It is critical when you think of networks to think about their dynamics. A lot of times, people fail to understand networks because they focus on the statics. They think about topology; they think about the architecture of the network. They think about how people are connected, which is of course incredibly important and not easy to understand, either. While on the one hand the topology can be understood or seen as fixed or existing, on the other hand this topology is itself mutable and changing and intriguing, and the origin of this topology and its change is itself a difficult thing.

But here is something else: Once you have recognized that there is a topology, the next thing you must understand is that there can be a contagion as well—a kind of process of flow through the network. Things move through it, and this has a different set of scientific underpinnings altogether. Understanding how things flow through the network is a different challenge from understanding how networks form or evolve. It is the difference between the formation and the operation of the network, or the difference between its structure and its function. Or, if you see the network as a kind of superorganism, it is the difference between the anatomy and the physiology of the superorganism, of the network. You need to understand both. And they both interconnect and affect each other, just as in our bodies our anatomy and our physiology are interrelated.

Nicholas A. Christakis

This is what James and I are tackling right now; we have started with several projects that seek to understand the processes of contagion, and we have also begun a body of work looking at the processes of network formation—how structure starts and why it changes. We have made some empirical discoveries about the nature of contagion within networks. And also, in the latter case, with respect to how networks arise, we imagine that the formation of networks obeys certain fundamental biological, genetic, physiological, sociological, and technological rules.

So we have been investigating both what causes networks to form and how networks operate. In terms of their operation, we have tackled some initial problems. For example, a few years ago, we became interested in the claim that there was an obesity epidemic. The word "epidemic" has a couple of meanings. First of all, it means that there is a higher prevalence now than in some previous time. It also includes the basic idea that there is something contagious that is spreading from person to person. There is no doubt that the prevalence of obesity is rising. What was not obvious to us was whether obesity could be seen as an epidemic in the other sense of the word. Was it spreading from person to person?

We wanted to study whether this was the case. Could obesity flow through networks? Could one person's body type actually influence the body type of others around him, and around them, and around them, in a cascade effect? People often take for granted that things can spread in a network, like fashions in clothes, but they were often surprised when we were able to show that obesity spreads in a network. How did we do that? We needed to come up with a source of data that contained information about people's position in a network, the architecture of their ties—who they knew and who those people knew and who those people knew and so forth. We also needed a source of data on people's weight and other information about them. And we needed it for a long period of time with repeated observations on these people. This was a difficult challenge. No data set to our knowledge existed before we made the one I am about to describe.

We hit upon the idea of working with a very well-known epidemiological study called the Framingham Heart Study, which was funded by the federal government and had been ongoing since 1948 in Framingham, Massachusetts, not far from Boston. In a basement there, we found a bunch of records in which the people who were responsible for tracking the thousands of participants kept information about how to reach the participants every two to four years so that they could come back for an examination and to fill in surveys and the like.

When we saw these paper records, it was immediately obvious that they contained valuable information, because they told us where the people lived, who their family members were, who friends of theirs were, where they worked, and so on. And it occurred to us that we could computerize these records, and that by dumb luck a lot of the people who were relatives or friends or neighbors of these individuals would also be participants in the heart study.

Therefore, we could reconstruct the social network ties of a sample of 12,000 people over the course of thirty-two years and have information about them that had been collected repeatedly across time. In so doing, we could set the stage for a set of analyses that looked at how weight gain in one individual spread from that individual and caused weight gain in other individuals, and how that in turn cascaded through the network. What we found when we did this study is that weight gain in your friends makes you gain weight, and weight gain among people beyond what we call your "social horizon" ripples through the network and affects you.

To us, it is a very, very fundamental observation that things happening in a social space beyond your vision—events that occur or choices that are made by people you don't know—can cascade in a conscious or subconscious way through a network and affect you. This is a very profound and fundamental observation about the operation of social life, which we initially examined while looking at obesity. We found that weight gain in a variety of kinds of people you might know affected your weight gain—weight gain in your friends, in your spouse, in your siblings, and so forth. Moreover, people beyond those to whom you were directly tied also influenced your weight, people up to three degrees removed from you in the network. And, incidentally, we found that weight loss obeys the same properties and spreads similarly through the network.

It is one thing to observe the spread of phenomena through the network; it is another to take the next step and begin to identify a mechanism of spread. In the case of obesity, we formulated a variety of ideas and were able to test some of them. And we have a variety of new experiments in mind to continue to investigate the spread of obesity and other phenomena.

One possible mechanism is very simple: biological contagion. There is a variety of work being done by biologists looking at viruses and bacteria that could spread from person to person and contribute to the obesity epidemic. Our work is completely consistent with that, but this is not what we are interested in.

We are interested not in biological contagion, but in social contagion. One possible mechanism is that I observe you and you begin to display certain behaviors that I then copy. For example, you might start running and then I might start running. Or you might invite me to go running with you. Or you might start eating certain fatty foods and I might start copying that behavior and eat fatty foods. Or you might take me with you to restaurants where I might eat fatty foods. What spreads from person to person is a behavior, and it is the behavior that we both might exhibit that then contributes to our changes in body size. So, the spread of behaviors from person to person might cause or underlie the spread of obesity.

A completely different mechanism would be for there to exist not a spread of behaviors, but a spread of norms. I look at the people around me and they are gaining weight. This changes my idea, consciously or subconsciously, about what is an acceptable body size. People around me who start gaining weight reset my expectations about what it means to be overweight or thin, and this is what spreads from person to person: a norm. It is a kind of meme (but it is not quite a meme) that goes from person to person.

In our empirical work so far, we have found substantial evidence for the latter mechanism, the spread of norms, more than the spread of behaviors. It is a bit technical, but I will explain it. In our empirical work on obesity, we found two lines of suggestive evidence for a spread of norms. The first line of evidence caught everyone's attention, and frankly it caught our attention when we noted it. It showed that it did not matter how far away your social contacts were; if they gained weight, it caused you to gain weight. This was the case whether your friend lived next door, 10 miles away, 100 miles away, or 1,000 miles away. Geographic distance did not matter to the obesity effect, the interpersonal effect.

Another finding from looking at the spread of smoking behavior was that if you stop smoking, it makes me stop smoking and there is a spread of smoking-cessation behavior, which itself is something we are investigating. Pertinent for the present purpose, however, is that, after taking into account the spread of smoking-cessation behavior, it did not efface the spread of obesity. In other words, accounting for one particular behavior, smoking cessation (which is known to increase weight at the individual level), did not undo the spread-of-obesity effect. This is an example in which it is not a spread of a behavior that causes the spread of obesity. This finding, coupled with the finding regarding the lack of decay with geographic distance, suggests to us that it is a norm rather than a behavior that is spreading.

Why? Because for a behavior to spread, typically, you and I would have to be together. We would have to go running together, share meals together, or copy each other's behavior in some way. And that should decay with geographic distance, because the farther away you are, the less time we can spend together. But a norm can fly through the ether. I might see you once a year and see that you have gained a tremendous amount of weight, which resets my idea about what an acceptable body size is. And minimal contact might be enough.

If I go see my brother Dimitri for Thanksgiving, no matter how much food we eat, no matter how much we share the behavior of eating, it will not change my weight that one day. But if I see him and he has gained a lot of weight, it can change my idea about what an acceptable body size is and, in that way, the spread of the norm can cause the spread of obesity.

Clothing fashions spread in our society. One way this can happen is you see people who reset your idea of what is fashionable. Another is more pragmatic. I take you shopping and we pick something out together. I say, "Oh, I heard about a new store," whatever. Those are two different ways in which fashions might spread.

We also have found in our work that things beyond obesity and smoking cessation spread in networks. Happiness spreads in networks. If your friend's friend becomes happy, it ripples through the network and can make you happy. We see clusters of happy and unhappy individuals in the social network like blinking lights in this complex fabric where some people are happy and some people are unhappy and there is a kind of gray zone between them. There is an ongoing kind of equilibrium that is reached in this social space. We have found that depression can spread, and drinking behaviors can spread, and the kinds of foods people choose to eat can spread (a taste for tastes can spread, as one of my graduate students is studying). All of this using the initial Framingham Heart Study social network dataset.

The spread of obesity occurs via a variety of mechanisms, but we find evidence at a minimum for the role of norms. How can it be that there is a role of norms in the spread of obesity when the ideology in our society regarding thinness is the same as it ever was? The supermodels are just as thin as they ever were. Interestingly, there has been some change in the weight status of celebrities (there were always overweight celebrities, but I think there may be more now than there used to be); but supermodels are certainly as thin as they ever have been.

This is the difference between ideology and norms. People see these images of supermodels, but they might be less influenced by them than by the actions and appearance of the people immediately around them. For example, we see that people might behave badly and engage in criminal acts. We still have the ideology that the Bill of Rights and the Constitution hold, and that there is goodness and there is evil. But people still behave badly when they are surrounded by people who behave badly. Again, it is the difference between norms and ideology, and this is how we square the circle in terms of why it is that there can be a spread of obesity, or an obesity epidemic, even though as a society we still seem to revere a kind of body type different from the one we are increasingly seeing.

Nicholas A. Christakis

James Fowler and I never expected to get the level of attention that we have for our work. On the morning of July 26, I knew we were going to be in the *New York Times*, because we had been interviewed by all these reporters prior to the appearance of our paper in the *New England Journal of Medicine*. I'd been in the newspaper before, and the work was in a prominent journal, so I thought I knew what to expect, but when I went out to my driveway that day, the article was unexpectedly on the front page of the *New York Times*. I went inside and said to my wife, "You're not going to believe this." And after that, it just did not stop. But what was interesting to me was that it wasn't just the *Times*—pretty much every newspaper thought this was something interesting. The coverage by the *Washington Post* and the *Chicago Tribune* was especially impressive. We had been working on the project for five years and we thought it was interesting, but we didn't think there would be so much popular interest.

Incidentally, we are not claiming that the fact that obesity might spread through social networks—or that the social network phenomenon might be relevant to the obesity epidemic—is the *only* explanation for the epidemic. No doubt there are many explanations. Those explanations, however, are not genetic. Our genes haven't changed in the past thirty years.

The real explanations for the obesity epidemic are exclusively socio-environmental—things having to do with the increasing consumption of calories in our society: Food is becoming cheaper, the composition of food is changing, there is increasing marketing of foodstuffs and the like. Also, clearly, there has been a change in the rate at which people burn calories due to an increase in sedentary lifestyles, the design of our suburbs, and a whole host of such explanations.

We are not claiming that such explanations are not relevant. No doubt they are all part of the obesity epidemic. We are just saying that networks have this fascinating property whereby they magnify whatever they are seeded with. And so if you get something like obesity going in a networked population, it can spread.

It should also be possible to trigger a spread in weight loss. We see this on a micro-dynamic scale in high schools, in niches of girls who start trying to compete with each other in terms of weight loss. One of the articles that came out about our work in the *Guardian* had pictures of the Spice Girls and the women in *Sex and the City* and talked about the "skinny flu" spreading from performer to performer. I think it was the first time Posh Spice and James Fowler were featured in the same paragraph.

So, you can get a kind of rush to the bottom, as well. In fact, after our work was published, we were contacted by a bunch of people who were seeking to treat people with eating disorders and who wondered if some of these network properties could be exploited clinically to improve the health of various individuals.

We also mention in our paper in the *New England Journal* the possible relevance of so-called mirror neurons, which is a mechanism that I didn't touch on earlier. One possibility besides biological contagion is that by watching you exhibit certain kinds of behaviors like eating or running, I start to copy those behaviors mentally in a mirror-neuron kind of way. And this facilitates my exhibiting the same behavior.

It is actually quite complicated to know how to exploit these network phenomena in a situation like the one we have been discussing, because if you have a lot of people of one body type and you introduce somebody of a different body type, it is unclear who will influence whom. The thin person might gain weight, or the overweight people might lose weight. Or both. It is a very com-

Nicholas A. Christakis

plicated dynamic, which again requires a kind of deployment of a certain kind of data and methods to begin to understand.

I should also stress something very important, which is that James Fowler's and my primary focus is not obesity, it is networks. Obesity happens to be an incredibly important public health problem and was something very important to study, above all because it showed how obesity was something that could spread in social networks, which people might not have realized. If we had shown, for example, that fashion spreads in social networks, that might be much less interesting to people. But if you can show that something like obesity or happiness or even goodness spreads in social networks, you are on new terrain.

Incidentally, some of these things also touch on very old philosophical and social scientific concerns, as I mentioned earlier, because they raise questions about free will. If my behaviors and my thoughts are determined not just by my own volition, but by the behaviors and thoughts of other people to whom I am connected—and are even determined by the behaviors and thoughts of other people whom I do not know and who are beyond my social horizon but connected to people to whom I am connected—it speaks to the issue of free will. Are my thinking and my behavior truly free, or are they constrained because I am part of a social network? To the extent that I am part of this human superorganism, does that reduce my individuality? And does this give us more or less insight into human behavior?

Because we are talking about networks of human beings rather than networks of neurons or computers, it is the case that I am not just plunked down in a network that is determined by some kind of exogenous physical law. There is no doubt that the topology obeys certain biological and psychological rules and laws, but it is also true that I can choose who my friends are and say, "You know, I don't like these friends; I am going to pick new friends."

That is, your desires and ideas can influence the structure of your network. For example, if you have ideas that foster certain kinds of ties, those ties in turn foster and support certain kinds of ideas. You can imagine a circumstance in which certain kinds of ideologies can survive and offer certain kinds of advantages because they bind the group together, or tear it apart, in particular kinds of ways. We have been thinking a little bit about this in terms of groups of people who seem to evince what would appear to be self-destructive behaviors, but our thoughts in this regard are still very preliminary.

Let's talk about our work with Facebook. The Framingham Heart Study network was something we had to painfully assemble using archival records about particular kinds of individuals. Even in the five years since we began to work on that project, the leaps in telecommunications and the emergence on the Internet of sites and technologies that are affirmatively organized as social networks—whereby people actually form and display their networks—have provided amazing research opportunities. When it comes to the Internet, we are no longer merely talking about networks of computers or networks of people who are in communication with each other, but we are talking about truly social networks, such as Facebook and Myspace and Friendster and LinkedIn.

The emergence of these technologies is a gold mine for social scientists in general, and certainly for people like James Fowler and myself, who are interested in social networks. We have begun a set of projects that exploit naturally occurring social networks on the Internet, like Facebook, or that seek to exploit the Internet to manipulate social networks in a variety of experimental ways—

Nicholas A. Christakis

for example, in some work I have been doing with Damon Centola and others.

Our Facebook project is only tangentially related to health, but is very much related to other concerns we have regarding the connection and contagion that take place with regard to the formation and operation of networks. We have been working at one particular university, where we have taken repeated cuts through the network. That is, a key feature is that we have longitudinal resolution across time and so can observe the network at several points in time, which prior generations of social scientists could not easily do.

We have trawled through this large social network and grabbed the information about people in the network and their social ties that is available on Facebook—information having to do with their tastes, with the people with whom they appear in photographs, and so on. For example, a person might have an average of 100 or 200 friends on Facebook, but they might only appear in photographs with ten of them. We would argue that appearing in a photograph constitutes a different kind of social tie than a mere nomination of friendship.

By exploiting these kinds of data and a variety of computer science technologies, we have been able to build a network that changes across time and to trace the flow of tastes through the network (for instance, how as I start listening to a particular kind of music, you start listening to a particular kind of music). We have been able to study homophilic properties—the idea that birds of a feather flock together. How and why do people form unions? Do they depend upon particular attributes, tastes, and the like? We have been able to study how these types of things—both the topology of the network and the things that flow through it—change over time.

In one project developed from this research, we considered whether someone wants to keep his or her information private on the Internet. Initially, without trivializing this serious topic, the issue of privacy was a methodological nuisance. But then we realized that, in addition to its conceptual importance, we could treat privacy as a taste. And we saw that the taste for privacy flowed through the network so that if I adopt privacy settings on Facebook, the people to whom I am connected will be more likely to adopt privacy settings.

So here we observe yet another phenomenon. We have talked about the flow of obesity through a network, we have talked about the flow of happiness through a network, we have talked about the flow of smoking cessation through a network, and we have talked about the flow of fashions through a network. Now we are talking about the flow of tastes in privacy through the network. And tastes in all kinds of other things, like music, movies, or books, or a taste in food. Or a flow of altruism through the network. All of these kinds of things can flow through social networks and obey certain rules we are seeking to discover.

Nicholas A. Christakis

12.

The Next Renaissance

Keynote Address at the Personal Democracy Forum

Douglas Rushkoff

Media analyst; documentary writer; author, Program or
Be Programmed

To me, "personal democracy" is an oxymoron. Democracy may be
a lot of things, but the last thing it should be is "personal." I understand "personal responsibility," such as a family having a recycling
bin in which they put their glass and metal every week. But even
then, a single recycling bin for a whole building or block would be
more efficient and appropriate.

Democracy is not personal, because if it's about anything, it's
not about the individual. Democracy is about others. It's about
transcending the self and acting collectively. Democracy is people
participating together to make the world a better place.

One of the essays in this conference's proceedings—the book
Rebooting America—remarks snarkily, "It's the network, stupid."
That may go over well with all of us digital folks, but it's not
true. It's not the network at all; it's the people. The network is
the tool the new medium that might help us get over the bias of
our broadcasting technologies. All those technologies that keep us
focused on ourselves as individuals, and away from our reality as
a collective.

This focus on the individual, and its false equation with democracy, began back in the Renaissance. The Renaissance brought us

wonderful innovations, such as perspective painting, scientific observation, and the printing press. But each of these innovations defined and celebrated individuality. Perspective painting celebrates the perspective of an individual on a scene. The scientific method showed how the real observations of an individual promote rational thought. The printing press gave individuals the opportunity to read, alone, and cogitate. Individuals formed perspectives, made observations, and formed opinions.

The individual we think of today was actually born in the Renaissance. The Vitruvian Man, Da Vinci's great drawing of a man in a perfect square and circle—independent and self-sufficient. This is the Renaissance ideal.

It was the birth of this thinking, individuated person that led to the ethos underlying the Enlightenment. Once we understood ourselves as individuals, we understood ourselves as having rights. *The Rights of Man.* A right to property. The right to personal freedom.

The Enlightenment—for all its greatness—was still oh-so-personal in its conception. The reader alone in his study, contemplating how his vote matters. One man, one vote. We fought revolutions for our individual rights as we understood them. There were mass actions, but these were masses of individuals fighting for their personal freedoms.

Ironically, with each leap toward individuality there was a corresponding increase in the power of central authorities. Remember, the Renaissance also brought us centralized currencies, chartered corporations, and nation-states. As individuals become concerned with their personal plights, their former power as a collective moves to central authorities. Local currencies, investments, and civic institutions dissolve as self-interest increases. The authority associated with them moves to the center and away from all those voting people.

Douglas Rushkoff

The media of the Renaissance—the printing press—is likewise terrific at mythmaking. At branding. Its stories are told to individuals, either through books or through broadcast media directed at each and every one of us. Its appeals are to the self and self-interest.

Consider any commercial for blue jeans. Its target audience is not a confident person who already has a girlfriend. The commercial communicates, Wear these jeans and you'll get to have sex. Who is the target for that message? An isolated, alienated person who does not have sex. The messaging targets the individual. If it's a mass medium, it targets many, many individuals.

Movements, like myths and brands, depend on this quality of top-down, Renaissance-style media. They are not genuinely collective at all, in that there's no promotion of interaction between the people in them. Instead, all the individuals relate to the hero, ideal, or mythology at the top. Movements are abstract—they have to be. They hover above the group, directing all attention toward themselves.

As I listen to people talk here—well-meaning progressives, no doubt—I can't help but hear the romantic, almost desperate desire to become part of a movement. To become part of something famous, like the Obama campaign. Maybe even get a good K Street job out of the connections we make here. It's a fantasy perpetrated by the TV show *The West Wing*. A myth that we want to be part of. But like any myth, it is a fantasy—and one almost entirely prefigured by Renaissance individualism.

The next renaissance (if there is one) is not about the individual at all, but about the networked group. The possibility for collective action. The technologies we're using—the biases of these media—cede central authority to decentralized groups. Instead of moving power to the center, they tend to move power to the edges.

Instead of creating value from the center—like a centrally issued currency—the network creates value from the periphery.

This means the way to participate is not simply to subscribe to an abstract, already-written myth, but to do real things. To take small actions in real ways. The glory is not in the belief system or the movement, but in the doing. It's not about getting someone elected, it's about removing the obstacles to real people doing what they need to do to get the job done. That's the opportunity of the networked, open-source era: to drop out of the myths and actually do.

Sadly, we tend to miss the great opportunities offered us by major shifts in media.

The first great renaissance in media, the invention of the alphabet, offered a tremendous leap for participatory democracy. Only priests could read and write hieroglyphs. The invention of the alphabet opened the door for people to read, or even possibly write, for themselves. In Torah myth, Moses goes off with his father-in-law to write the laws by which an enslaved people could now live. Instead of simply accepting legislation and government as a preexisting condition—the god Pharaoh—people would develop and write down the law as they wanted it. Even the Torah is written in the form of a contract, and God creates the world with a word.

Access to written language was to change a world of blind, enslaved rule-followers into a civilization of literate people. (This is what is meant when God tells Abraham, "You will be a nation of priests." It means they are to be a nation of people who transcend hieroglyphs or "priestly writing" to become literate.)

But this isn't what happened. People didn't read the Torah—they listened as their leaders read it to them. Hearing was a step up from simply following, but the promise of the new medium had not been seized.

Likewise, the invention of the printing press did not lead to a civilization of writers—it developed a culture of readers. Gentlemen sat reading books, while the printing presses were accessed by those with the money or power to use them. The people remained one step behind the technology. Broadcast radio and television are really just an extension of the printing press: expensive, one-to-many media that promote the mass distribution of the stories and ideas of a small elite.

Computers and networks finally offer us the ability to write. And we do write with them. Everyone is a blogger now. Citizen bloggers and YouTubers who believe we have now embraced a new "personal" democracy. Personal, because we can sit safely at home with our laptops and type our way to freedom.

But writing is not the capability being offered us by these tools at all. The capability is programming—which almost none of us really know how to do. We simply use the programs that have been made for us, and enter our blog text in the appropriate box on the screen. Nothing against the strides made by citizen bloggers and journalists, but big deal. Let them eat blog.

At the very least, the opportunity here is not to write about politics or—more likely—comment on what someone else has said about politics. The opportunity is to rewrite the very rules by which democracy is implemented. The opportunity of a renaissance in programming is to reconfigure the process through which democracy occurs.

If Obama is indeed elected—the first truly Internet-enabled candidate—we should take him at his word. He does not offer himself as the agent of change, but as an advocate of the change that could be enacted by people. It is not for government to create solar power, for example, but to get out of the way of all those people who are ready to implement solar power themselves. Responding to the willingness of people to act, he can remove regulations to restrict its proliferation developed on behalf of the oil industry.

In an era when people have the ability to reprogram their reality, the job of leaders is to help facilitate this activity by tweaking legislation, or by supporting their efforts through better incentives or access to the necessary tools and capital. Change does not come from the top—but from the periphery. Not from a leader or a myth inspiring individuals to consent to it, but from people working to manifest it together.

Open source democracy—which I wrote about a decade ago—is not simply a way to get candidates elected to office. It is a collective reprogramming of the social software, a disengagement from the myths through which we abdicate responsibility, and a reclamation of our role as citizens who participate in the creation of the society in which we want to live.

This is not personal democracy at all, but a collective and participatory democracy where we finally accept our roles as the fully literate and engaged adults who can make this happen.

Douglas Rushkoff

13.
Digital Power and Its Discontents

Evgeny Morozov and Clay Shirky

Evgeny Morozov: *Commentator on the Internet and politics; contributing editor,* Foreign Policy; *author,* The Net Delusion: The Dark Side of Internet Freedom

Clay Shirky: *Social and technology network topology researcher; adjunct professor, New York University Graduate School, Interactive Telecommunications Program (ITP); author,* Cognitive Surplus

> The dreams of network utopians vs. the realists. Is the Internet a medium of emancipation and revolution—or a tool of control and repression? Did Twitter and Facebook stoke the flames of rebellion in Iran, or did they help the authorities unmask the rebels?
>
> —*Frankfurter Allgemeine Zeitung*

CLAY SHIRKY: Evgeny, I think this may be a frustrating hour, because I think you and I disagree with each other less than you disagree with a lot of the people you're calling Internet utopians. For instance, you recently picked on the John Perry Barlow piece "A Declaration of the Independence of Cyberspace," which is so over-the-top libertarian, presenting cyberspace as a separate sphere unconnected to the rest of the planet, that it didn't really have much effect on practical matters like foreign policy.

EVGENY MOROZOV: I guess we're talking about a recent essay of mine that appeared in the *Wall Street Journal* in February 2009. It was published a week after yet another overhyped wave of Iranian protests came to nothing. But this time something was different in how that failure was explained in the media. Suddenly, I could sense some public frustration—even in the *New York Times*—about how the Internet could have actually thwarted the protests, making them more disorganized. That's something I really wanted to play with in that essay. But since the *Wall Street Journal* wanted me to offer a critique of techno-utopianism, I had to venture beyond recent events and see the kinds of ideas guiding governments in this space. Thus, the real objective was not to pick on John Perry Barlow—who in 1996 wrote "A Declaration of the Independence of Cyberspace," which is one of the seminal texts of cyber-libertarianism—or any of the other early thinkers. It was more to reveal that we are currently facing a huge intellectual void with regard to the Internet's impact on global politics.

But the lack of a coherent framework does not really prevent us from embracing the power of the Internet. There is certainly a lot of excitement within governments—both democratic and authoritarian ones—about using the Internet to advance their political agendas, both at home and abroad. The kind of assumptions that politicians need in order to decide their policies all have to come from somewhere. And much of what has been said about the Internet in the past seems intellectually invalid today. Still, most of the assumptions made by politicians seem to be rooted in early cyber-libertarian discourses about the Internet and politics. A lot of those early discourses took shape in particular (and very different) contexts. If you look at John Perry Barlow's declaration, it was produced in the context of attempts to regulate the Internet

Evgeny Morozov and Clay Shirky

in America, in 1996. It had nothing to do with Iran and only very little with the world outside of the U.S. We do need a new theory to guide us through all of this, for the old theories are no good.

SHIRKY: Yes, I agree with that, and with regard to "A Declaration of the Independence of Cyberspace"—ten years ago I was teaching that at NYU classes as an example of sloppy political thinking, so I think we've known that those theories were no good for a while now.

You end your *Journal* essay with a fairly evocative paragraph saying, "The State Department can't abandon ideas of trying to harness the Internet for democracy, but it should come up with a policy that's more in line with what's possible, or what works." If you could give them a piece of advice—and let's put it in an absolutely specific context—if you could give Secretary Clinton, Alec Ross, and Jared Cohen advice about using the Internet to further U.S. foreign policy goals, what would you say?

MOROZOV: "Do no harm" would be my first operating principle. By building very close and heavily publicized alliances with Google, Twitter, and any other big technology companies, officials at the U.S. State Department are presenting these companies as if they were some kind of Web 2.0–era reincarnation of the Radio Free Europe—"Radio Free Internet," if you will—which they aren't. These companies have their own commercial agendas; they're primarily interested in making money, not in spreading American ideals. Yes, the Internet may help Google to sell more ads *and* advance American interests at the same time, but it's not exactly how the State Department operates. At least in theory, we're not promoting Internet freedom because it helps America to sell more books, films, and newspapers (buying stuff—like oil—is a whole different story!). We're promoting Internet freedom for freedom's own good. So the real question is how to leverage the undeniable power of these companies without presenting them as extensions of U.S. foreign policy.

Thus, when someone from the State Department takes executives from Google or Twitter to tour around the world, to go to Siberia, it just looks ridiculous and plain wrong to me. It makes people question this sudden proximity that exists between politicians and companies, especially when Google is also cooperating with the National Security Agency. If I were working for any government, authoritarian or democratic, I would not be very happy with the majority of my citizens doing their e-mail with a company that has secret dealings with the likes of the NSA. That's a legitimate concern. Do we really want Google to be seen as the next Halliburton?

SHIRKY: To that commercial question, I'm struck by the parallel with the U.S. pavilion in 1959 in Moscow, where the government built, among other things, the U.S. kitchen (the site of the "Kitchen Debate" between Khrushchev and Nixon), and that was an explicitly commercial projection of American identity. From my point of view the Kitchen Debate was a seminal moment in shifting the terms of the debate. I'd ask what's the difference between Jared Cohen of the State Department taking Eric Schmidt of Google or Jack Dorsey of Twitter to North Africa, or Siberia, or what have you, and the U.S. helping General Electric get into Moscow in 1959?

MOROZOV: At that time no one was expecting that people would start using kitchen appliances to overthrow their government, right?

SHIRKY: You mean the "toaster revolution" . . .

MOROZOV: Something like the "toaster revolution" actually did happen in Iran last summer—only without the commercial elements. There was a viral offline campaign asking people to turn on all their electronic appliances at a set time in order to shut down the power grid. This was a human "denial of service" attack.

But let's not get carried away with the Khrushchev analogy. There is definitely a greater level of politicization attached to the use of Twitter, Google, and Facebook in authoritarian conditions. People who are now using Twitter in Iran are marked as potential enemies of the state, much like those who are using proxy servers in order to access banned content. You may be using it to download pornography, but as far as the state is concerned, you'll be seen as a potential political enemy anyway.

SHIRKY: The Facebook example is interesting, because unlike a lot of the conversation about Twitter, mobile phones, proxy servers, and so forth, the Iranian government blocked Facebook before the election, they blocked it on June 8 or 9, and the election happened on June 12. So nobody knew what was coming after the election, and Iran still shut Facebook down.

Do you think the Iranian government overreacted to Facebook? Is your thesis that Facebook and Twitter are not, in fact, terribly effective political tools, and the Iranian government overreacted? Or do you believe those tools were effective in June, but the Iranian government responded in such a way that they are no longer effective?

MOROZOV: Facebook is a very particular example when it comes to Iran because there were blockages and unblockages throughout the election campaign. If you closely study it from January 2009 to June 2009, there have been multiple instances when it was blocked, and then unblocked, and then blocked again. But, to me, the fact that they blocked Facebook doesn't mean anything. All it means is that they could block Facebook—and they did. The fact that they are blocking does not necessarily endow Facebook with some special political meaning. Look at other countries, such as Cambodia in 2007, when they had elections and imposed the so-called tranquility period wherein all mobile operators agreed to turn off all text-messaging services for three days during the election period. Were they expecting a text-message rebellion? I don't think so. The point here is that they could do it and they did.

SHIRKY: As did Singapore with respect to blogs. But this seems to be an indication of real political fear. Singapore and Cambodia framed this censorship as connected to a positive set of political values, where the political rationale was that the citizens would be better off with this tranquil period to reflect on who the better leader should be rather than, God forbid, talking to their friends and neighbors. Now, I don't buy these rationales, but even if you do frame censorship that way, the censorship seems to me to be an explicitly political act.

When I see Cambodia, or Singapore, or Iran, shutting down a service that increases social coordination, my response is essentially the one Habermas proposed in *Structural Transformation of the Public Sphere*, which is that those regimes are trying, quite specifically, to dampen the public sphere. If that's a political judgment those governments are making, the question I am asking myself is: Are those regimes right in fearing better social coordination among the public?

I think the answer is yes. The Burmese example of communications use during their political struggle, followed by panicked shutdown, or the Ukrainian example from the Orange Revolution, or the successful Moldovan protests of last year, suggest to me that conditions under which a public that can self-identify and self-synchronize, even among a relatively small elite, is in fact a threat to the state.

This is one of the things I want to understand about your

videos, because while you and I are not polar opposites, we obviously have very different points of view about this. Do you believe that the synchronizing effect among a politically engaged public is (a) possible and (b) political, and if it is, what should the U.S. reaction to that be?

MOROZOV: First of all, there is symbolic value attached to censorship: It does help the Iranian government to signal to the rest of the world that they are still in charge. What the authorities would love to do is for everyone to believe that they are succeeding in their attempts to block Facebook (even if this is not, strictly speaking, the case). They would even love to issue a press release to that effect: "Yes, we're blocking Facebook because we are still in charge; we can do that, and we'll do that."

But if we look beyond just symbolic benefits that governments can derive from propaganda, I would argue that one of the reasons why the Iranian authorities have been so seemingly ineffective at blocking the Web is that they—correctly, in my view—also see tremendous value in watching anti-government Iranians coordinate their actions—publicly, mind you—on Facebook and Twitter. For example, they might be learning about the kind of groups and threats that are emerging. This intelligence value is something that we often tend to forget.

Second, I'm not sure that the synchronicity that Habermas talked about was measured in days, hours, and tweets as opposed to decades, centuries, and books. But that aside, was this new hyper-synchronicity actually present in the Iranian online campaigning, and how did it influence the actual protests? Yes, it was a very vibrant online campaign, but I didn't see it extending into real-world coordination all that much. How many completely uninitiated protesters have actually engaged with the real world because of something they had read on Twitter or Facebook? While there was synchronicity of online actions, I'm not sure that it translated well into coordinated protests in the streets.

SHIRKY: I don't think it could translate well into coordinated protests. To my eye, those Tehrani protests didn't look strongly directed. They looked more like effusions than planned events. But if I wanted to pick a maximum case for online coordination changing real-world politics, it would be in the role of women.

To take one example, the South Korean protests of 2008, following the importing of American beef after our mad-cow contamination, hinged on the ability of women to come forward rhetorically, particularly in a political environment where they're quite restricted physically and publicly. Then there's Neda Agha-Soltan, the famous martyr of the early events in Iran—that again strikes me as something that probably wouldn't have happened without these tools. Again, there's a small number of such political events to reason from, but it seems to me unlikely that the presence of the women in the protest movement would have happened without the social media having a coordinating effect.

MOROZOV: I'm not an expert on Iranian women, but from what I understand they have been experimenting with social media for at least a decade. So again, why was there so much activity on social media sites in Iran? Well, because so many people had access to it. From this perspective, most social media activity is just epiphenomenal: It happens because everyone has a mobile phone. But still: Despite millions of angry tweets and cell-phone cameras pointed at their faces, the Iranian government still cracked down on the protesters. Just look at what's happened in Iran over the past nine months: the fading protests, the growing division in the country. A lot of people had to emigrate, a lot of people were imprisoned, a lot of people were killed. As far as the political situation on the ground is concerned, it's quite grim: If there are some big positive developments, I don't see any.

I just don't see where there's very positive stuff that I'm missing. The brutal people would be there without social media.

SHIRKY: Concerning our debate in *Prospect* in December, I admitted to being a rhetorical bad actor in this conversation up until now. Since my area of concentration is social coordination among otherwise uncoordinated groups, I wasn't offering accounting for the full political sphere, I was only offering accounting for those groups, and you rightly pointed out that hierarchically managed groups have access to decisive action in a way that uncoordinated groups don't.

With that caveat, there do still seem to be effects that previously uncoordinated members of the social sphere are now having. I believe that the Green Uprising altered the balance between the theocracy and the military aspects of the Iranian government. Iranian political power moved in the direction of the military in the 2005 election, after the scare from Khatami's moderate government, and it seems to me that one of the effects, paradoxical and sad, of the Green Uprising is that it has pushed Iran into being an essentially military power.

The theocracy as a moderating force over popular passions was broken under the weight of the regime's inability to do anything in public without rekindling the uprising. They couldn't even keep the anniversary of the revolution from turning into antigovernment protest, and the way that they finally held down the Green movement in February was to amass an unbelievable number of brutal-minded cops.

One of the ways in which I might be wrong about the moderating effects of the political sphere would be if digital coordination does, indeed, create a movement of these states, where those that have some kind of political sphere emerging become more like Cambodia under the Khmer Rouge than like South Korea under the military dictatorship in the 1980s. Again, a small number of cases to reason from—Iran is just one example, and the uprising has been going on less than a year—but one of the open questions is whether or not the presence of a more engaged public will actually make governments more brutal rather than more open to change.

MOROZOV: One of the reasons I've been so unhappy with how the media have been covering the role of the Internet in Iran—and this I guess also has to do with them reading certain things into your book that you did not intend to say there—is the almost exclusive focus on analyzing what the Internet has done to protest movements, at the expense of thinking about its impact on everything else. But if we focus only on how people coordinate themselves with the help of social media before, during, or after the elections, we miss many other effects that the Internet is having in public, social, and political life in authoritarian states, especially in the long term.

Shouldn't we also be asking whether it's making people more receptive to nationalism? Or whether it might be promoting a certain (hedonism-based) ideology that may actually push them further away from any meaningful engagement in politics? Does it actually empower certain nonstate forces within authoritarian states that may not necessarily be conducive to democracy and freedom? Those are all big questions which we cannot answer if we just focus on who gets empowered during the protests, the state or the protesters, because some countries, well, don't have that many protests. Or elections. China doesn't have national elections.

Evgeny Morozov and Clay Shirky

SHIRKY: Well, they do at the local level, and that's where we're seeing more protest. Sichuan is a great danger to the Chinese central party because that's where—to your point about modernizing—they're desperate to modernize local and regional political representation of economies. If the dead hand of the state isn't removed at the level of productive factories, they can't continue to generate the growth they need. However, that requires the kind of engagement that has historically been connected to greater political demands made of the state.

MOROZOV: It's all a matter of questions that we want to ask. If the question we are asking is, How does the Internet impact the chances for democratization in a country like China?, we have to look beyond what it does to citizens' ability to communicate with one another or their supporters in the West. I recently found a very fascinating set of statistics: Apparently, by 2003 the Chinese government had spent $120 billion on e-government and something like $70 million on the Golden Shield, the censorship project. You compare those two numbers—$120 billion on e-government and $70 million on censorship—and you can sense that the Chinese are really excited by e-government. No surprises there: It can make their government more efficient, making it seem more transparent and resistant to corruption. This would only strengthen the government's legitimacy. Will it modernize the Chinese Communist Party? It will. Will it result in the establishment of democratic institutions that we expect in liberal democracies? It may not. If we want to know whether China is moving closer to embracing fully functioning democratic institutions and what kind of role the Internet would play in this process, there are no easy clear-cut answers here.

SHIRKY: This is one of the really interesting things about these questions, which is that you very quickly get a kind of philosophic vertigo. You think you're asking a question about Twitter, and suddenly you realize you're asking a question about, say, Hayek and markets. My bias is that nondemocratic governments are lousy at managing market economies over the long haul. That's a baseline assumption, and it affects the context of digital publics.

With that assumption as background, one of the questions you could ask is, How much is political sensitivity of the regime titrated to the price of oil? If oil goes back above $100 a barrel, the Iranian regime can do anything they like. They could destroy the intelligentsia in all of Tehran and still rule the country because they'd have so much cash from oil. If it goes under $50 and stays under $50, on the other hand, their ability to hold down populist uprising will be severely compromised.

MOROZOV: Whatever the bias, the truth is that we did have revolutions before Twitter.

SHIRKY: Yes, of course.

MOROZOV: And we did support those forces somehow, whether it was by smuggling technology, which did happen in Poland, smuggling in those Xerox machines, or just by making sure that the Polish political dissidents could link up with the Catholic Church.

SHIRKY: But smuggling the Xerox machines! That's exactly about driving the communications piece into the equation.

MOROZOV: Yes. But if you look at some emerging intellectual discussions that are now happening in "transition studies"— particularly among academics who study the causes of the 1989 revolution—you'll see quite a lot of disagreement about the reasons why communism collapsed. There are more and more revisionist voices—people like Stephen Kotkin, for example—who argue that the reason why communism collapsed was because its elites badly mismanaged the situation and the governments simply imploded from within. That's Kotkin's *Uncivil Society* thesis:

Communist governments just ran out of money and resources and couldn't support themselves, so whatever was happening at the grassroots level—with or without Xerox machines—didn't matter all that much. This, of course, overstates the case, but I think Kotkin is asking some important questions. You probably see the implications of his argument to the role of the smuggled Xerox machines: They may not have been all that important, for it was the fundamental economic unsustainability of communism that precipitated its collapse. So how many tweets are now being smuggled into Iran may not really matter in the long run.

SHIRKY: So I'll make an argument for why it is going to matter, which relates to the Iranian government's announced plan to ban Google's mail service and replace it with a "national e-mail service." I don't believe that the Iranian government will be able to run a good replacement for Gmail, not because of the censorship, but because I don't think they have talented enough systems administrators. I don't think they can keep the technology up. If they suddenly become tech support for their own country, that actually shifts the economy into a less productive mode.

Now, they can afford to give up half a percent a year GDP from their market economy if oil goes above $100 a barrel, but if oil stays at $70 and below, they can't afford to shave off that kind of growth. This is the idea of Iran becoming "temporary Burma" that you and I have been arguing about. I think Twitter, Facebook, etc., have pushed Iran into a serious compromise where they're willing to weaken their own communications infrastructure, a.k.a., shave tenths of a percentage point off GDP, to try to control the insurrection. That's understandable from their point of view, but still seems to me like a dangerous move in the longer term.

MOROZOV: But you are not proposing that lest they ban Gmail, they are going to crumble?

SHIRKY: No, no, no. I'm saying that their ban on Gmail indicates a fear of their own citizens communicating freely with one another, and I think that fear is justified.

MOROZOV: I think one problem with your analysis is that it takes almost everything that Iranian authorities say at face value. But it can all be interpreted differently: What if they just wanted to score propaganda points? To that effect, they announced a plan which they knew would never be executed. I lived in Belarus and saw enough crazy but completely meaningless threats and announcements by the state; usually, they went nowhere—they were meant as exercises in propaganda. If you look at China or Russia, much of their publicity is now run by Western PR firms who know how Western media works and know how to make it produce the coveted coverage. Why do we worry about American firms selling China technology that can then be used for censorship purposes rather than about, say, the PR and lobbying firms who cater to the publicity needs of authoritarian governments? I guess what I am trying to say is that these governments' media strategies are much more sophisticated and media-conscious than we take them to be. When we in the West are trying to second-guess what the Iranians actually meant, it does remind me of Kremlinology.

SHIRKY: It's a lot like Kremlinology, yes. So let me pose the question as a hypothetical rather than arguing about the face-value statements. You and I had a discussion in December, before the thirty-first anniversary of the Iranian Revolution and before the Gmail announcement, in which I posited that Iran was acquiring a kind of technological autoimmune disease. They are attacking their own communications infrastructure as the only way to root out the coordination among the insurrectionists.

You replied that that kind of communications blackout can be geographically limited and temporary; you can turn it on and turn

Evgeny Morozov and Clay Shirky

it off. The announcement of the Gmail ban seemed to me to be a national and nontemporary attempt to do the same thing to their communication infrastructure. So, were Iran to shut down parts of that infrastructure, do you believe that that starts to ramify in the economy in a way that mattered more than the Polish Xerox machines?

MOROZOV: No. You know, again, it depends on what exactly they'll be blocking. I don't think their ban on all e-mail exchanges—except those facilitated by a national provider—is ever going to happen. You have to carefully study the geopolitical context in which that threat had been made public. The Iranian authorities made the statement a week after Google had announced that they were talking with the NSA. That was a very propitious propaganda moment for the Iranian government to jump in and say, "We absolutely want to make sure that our citizens are not being watched by the NSA." Bingo: That's what they did. Masterful domestic propaganda once again.

SHIRKY: Changing the subject, one of the places I think the debate has gone awry is in overestimating the importance of the value of the access to information, and in underestimating the importance of the access of value to people. This is a mistake that dates from the dawn of the Internet.

In fact, if we could lower the censorship barriers between the West and China, could just remove the Golden Shield altogether while the Chinese retained the same degree of control over citizens and citizen communication, not much would change. If the Golden Shield stays up in its full form, but the citizen communication and coordination gets better, a lot will change. You could see evidence of this after the Shanghai earthquake.

MOROZOV: But my question is: In what direction—good or bad—would all of this change?

Digital Power and Its Discontents

SHIRKY: Well, right. As Robert Putnam has pointed out, social capital creates value for people inside the network while exporting harms to people outside the network.

I don't believe that freedom to communicate automatically brings in pro-Western governments, which is to say I am pro-democracy full stop, even if they are illiberal democracies like those Zakaria describes. I accept that there will be national movements whose goals are inimical to the foreign policy objectives of the West, but as long as those countries are democracies, I am less worried, frankly.

MOROZOV: Yes, but what comes first, democracy or Internet-based contention? It's not like once you have a new democracy, it's guaranteed to stay that way forever. If only democratization were that simple. Newly formed democracies are at their most vulnerable during the transition period. That's when they need a strong state to carry out the painful economic development and a broader program of liberalization. If you have a weak state entering a transition period—and it's fair to say the Internet would mobilize the groups that would make a weak state even weaker—chances are you would not end up with a democracy in the end. That's more or less what happened in Russia in the nineties.

SHIRKY: Let's get back to China. Here is what I think it looks like in China. They had the 2008 quake in Sichuan, the BBC finds out about it on Twitter, and the Chinese government finds out about it from QQ. The last time there was a quake of that magnitude, it took the Chinese three months to admit that it had happened. Here they don't even have a choice because the world is already reporting on it as they're kind of mobilizing.

This happened when they were having one of their "happy, happy, joy, joy" moments with the press, and they let the press report this whole thing, and then the mothers in Sichuan, a pretty

Evgeny Morozov and Clay Shirky

sympathetic group who have lost their children because the school buildings collapsed in the quake, realize that shoddy construction has caused the building to collapse. And all of a sudden they're protesting in public daily, and documenting those protests, and putting it up on QQ, and that's the first time the Chinese government has faced a radicalized population who had no preexisting coordination.

The only commonality to those women protesting was that they were mothers of school-aged children killed by an earthquake that brought the government-built buildings down. The degree, the sharpness of the eventual government crackdown on those protesters was so extraordinary that it suggested to me (a) that the government was not just worried but terrified and (b) that they were right to be terrified. And that's where the threat comes from in China—local politics, not national politics. There is going to be some breakaway internal province that's going to freak them out, and that is going to be the axis of change.

MOROZOV: I do agree that the ability of the Chinese government to control information flows has been somewhat—in some particular cases, quite significantly—eroded. But will they be able to adapt to this environment by engaging in new ways of propaganda? By selectively manipulating who gets to cover which story? Maybe. We do see evidence that this is what's happening. Last year we saw it in China's Yunnan Province, where a young man died in police custody. Instead of censoring thousands of comments that were gathering on sites like QQ, they let netizens blow off their steam. They solicited applications from them to become "netizen investigators" and eventually chose fifteen people who were then dispatched to examine the prison in question. They couldn't find anything, and wrote a very inconclusive report. That diffused the growing tensions—mind you, without need for any formal censorship (only later did netizens find out virtually all fifteen "investigators" were actually current or former employees of the state media). The moral of the story is that we tend to underestimate the government's ability to react to some of these news stories in ways that do not undermine their legitimacy and authority to the extent that we expect those two to be undermined.

SHIRKY: Let's talk about Belarus, because you're a Belarusian, and I linked the Susanne Lohmann work on information cascades to the flash-mob kids in Minsk in 2006, and that was a protest that failed. And you said something evocative in *Prospect* about that failure: The fence-sitters saw what was going on, and sensibly climbed higher on the fence. What do you think caused them to fail? What happened there? Because that's a real-world example of a protest movement that used social tools to coalesce, created a public stir, and then vanished.

MOROZOV: Not sure there was any kind of "public stir" in Belarus in 2006. One of the reasons why protests happened in the first place had to do with the fact that, yes, there were presidential elections, and one of the candidates in those elections was actually imprisoned shortly after the elections. It had nothing to do with social media. You know, if people had no Internet, they still would have showed up in the Square in the numbers that they did, probably. So, to me, the case of Belarus is even more unambiguous than the case of Iran: Social media didn't really play any role whatsoever in generating protests in the streets. And most people actually didn't show up. It was a relatively small protest, in part because the government is relatively popular.

But back to your question: My biggest problem with these flash-mobbing kids in Belarus was that they erroneously thought that the Internet presents an entirely new way of doing politics. They thought that they would build up and operate a fully virtual movement, that they would not need to bother with the dirty and

Evgeny Morozov and Clay Shirky

bloody business of opposing a dictator, a business that often entails harassments of all kinds, as well as bloodshed, intimidation, expulsion from universities. Let's not kid ourselves: That's what being in an opposition movement in an authoritarian country entails. It's never a pretty picture. So I do fear that some of these kids thought that the Internet offered a nice shortcut that would allow them to meaningfully challenge the dictator without having to go through any of that unpleasant stuff. They thought they could just blog the dictatorship away. I even know why some of them had such high hopes for virtual politics: It promised a viable alternative to the otherwise moribund oppositional politics of the country. In the particular case of Belarus, the country simply has a terrible, disorganized, always squabbling, and extremely unappealing opposition. No wonder so many smart young people do not want to be part of it. But the Internet presents them with a false choice. The reality is that they can either join and reshape this opposition from within, perhaps even using the Internet—or stay on the sidelines and get lost in free and abundant online entertainment.

SHIRKY: It seems to me that one of the things that the Lukashenko government, the Belarusian government, enjoys now is that nobody in the West really cares anymore what happens to the remaining authoritarian states because it's no longer the salient geopolitical question. Now it's just a little corner of the world that may have an additional spot of oil. Do you think those two things are correlated, which is to say increased interest in electronic communications has decreased particularly the U.S.'s willingness to exert pressure on Lukashenko, or do you think those two things just happened at the same time?

MOROZOV: Again, I think Belarus is not a good example, because the regime does enjoy popular support. But you look at Iran, you look at China, those are very much focal points of interest for the U.S. government. And yet the Iranian police were still cracking down on protesters, killing people despite the fact that everyone was armed with mobile phones. Could they have killed more? Probably. But I didn't see technology as a very effective deterrent. Neda was still killed despite the fact that there were people taking those videos.

But my concerns also have to do with how the Internet is changing the nature of political opposition under authoritarianism. I don't know if you've read Kierkegaard, but there are quite a few subtle undertones of Kierkegaard in my critique of Twitter-based activism. Kierkegaard happened to live during the very times that were celebrated by Habermas: cafes and newspapers were on the rise all over Europe, a new democratized public sphere was emerging. But Kierkegaard was growing increasingly concerned that there were too many opinions flowing around, that it was too easy to rally people behind an infinite number of shallow causes, that no one had strong commitment to anything. There was nothing that people could die for. Ironically, this is also one of my problems with the promiscuous nature of online activism: It cheapens our commitment to political and social causes that matter and demand constant sacrifice.

SHIRKY: One of the funny things that Habermas says in *The Structural Transformation of the Public Sphere* (from a not-long list of funny things) is that newspapers were best at supporting the public sphere when freedom of speech was illegal, so that to run a newspaper was an act of public defiance. Similarly, a protest which is relatively easy to coordinate at relatively low risk is not only less of a protest, but potentially draws off some of the energy that could go elsewhere.

MOROZOV: I am also not sure that bloggers make for great symbols of antigovernment campaigns. The kind of ordinary apolitical

Evgeny Morozov and Clay Shirky

people whom we are talking about—those who eventually muster up the courage to go and defy authorities in the streets—need to be led by people who are ready to take a brave stand, to sacrifice themselves, to go to prison, and to become the next Havels, Sakharovs, or Solzhenitsyns.

SHIRKY: The question is, Does a movement need a martyr, does it need an intellectual focal point who's willing to take a hit in order to make the point? And the second question is, Does that have to be one person? You know, the martyrization of Neda happened after the fact. We had no idea what she thought or meant to do. She could have been out on the street with her friends because this seemed like an interesting and important moment, but with no thought for risk, much less fatal risk. And so is that enough? Is it enough to have a Kent State, or does it have to be a Sakharov, which is to say someone who intentionally puts themselves in harm's way before harm comes their way?

MOROZOV: Well, I do think that the mass protest needs a charismatic leader—i.e., a Sakharov—to truly realize its potential. My fear is that a Solzhenitsyn would not be possible in the age of Twitter. He would probably end up in prison much sooner—and for a much longer period—than he actually did. I am not sure that Twitter would help him become a stronger and more charismatic public figure or gain the courage to write the first page of his book.

SHIRKY: I think that the thing that's worth fighting for here is the ability of citizens within a country to communicate with one another, and I remain convinced that that will have political ramifications. I don't think we just sit around waiting for the economies to fall apart. I think that to the degree that citizens can communicate with each other, that actually matters much more than access to information, or than communication with the outside world. What we should be worrying about is freedom of speech, lowercase *f*, lowercase *s*, not as a political right, but as just a daily capability. To what degree is that flourishing? Because in countries where that flourishes I think that the world would be better off even if non-Western norms develop in the population so served.

MOROZOV: Sure. I agree with most of what you've suggested. That said, there are still a lot of small things that the State Department needs to do that they're not currently doing very well. Take the sanctions on American companies that were supposedly lifted in early March. Sounds good, doesn't it? But then you look at the matter a bit closer and it turns out that if you are an Iranian living in the U.S. and you want to have a simple Google ad displayed on your website and write it in Farsi, thus earning some money and obviating the need to beg foreign governments for funding, you still can't do it.

And there are a lot of very specific granular things like this one that can be done. At this point, I think the State Department is literally like an elephant in a china store (only the "China" in this particular case happens to be capitalized). Too bad: Now is the worst possible time for them to be an elephant, for this is how the U.S. will always be remembered. If the Iranians, the Chinese, and the Russians get the impression that Silicon Valley is in bed with the State Department, that impression is likely to persist for quite some time, maybe forever (once again: try convincing foreigners that oil companies don't control Washington). Just like most foreign publics developed an impression—thanks to eight years of Bush—that the U.S. promoting democracy necessarily entails regime change, they may soon develop similar impressions about "Internet freedom." So I think the diplomats just have to be very careful, and focus on ironing out these micro problems instead of saying that, yes, we've developed this partnership with Twitter and everyone should know about it! It's the kind of "public diplomacy" that begs for being less "public."

14.

Does Technology Evolve?

W. Brian Arthur

Citibank Professor, Santa Fe Institute; author, The Nature of
Technology: What It Is and How It Evolves

In my career I have looked at very disparate subjects or areas of
interest. I got a chair at Stanford very early on by being an expert
in human fertility in the Third World, believe it or not, as a de-
mographer. I've been interested in the economy; I've been interested in certain types of mathematics; I've been interested lately
in technology. I'm getting to an age now where I can start to look
back and think, What on earth was all that about? What was the
common thread? I realize there is a common thread and it's very
deep inside me.

One of the things I was interested in also was what became
known later as complexity and all the stuff that the Santa Fe Institute became known for. It turned out I got to Stanford at the
right time, in the mid-eighties, when a lot of that was unfolding.
So I thought: demography, complexity, economics, technology. I
began to think in a rather despondent way that maybe I was a dil-
ettante. I dabbled here and dabbled there and there wasn't really
a common theme. But in the last year or two I have realized there
was a common theme. This is really what motivates me. I'm interested in systems, the unfolding or the development of systems or
of patterns.

When I studied at Berkeley—this was late sixties/early seventies—I wasn't interested in what became known as equilibrium economics, where everything is static and unmoving; I specialized in the economics of development. Then, when I got interested in complexity, I realized it was about systems, that complexity is basically about systems that are reacting to the overall situation they create, individual elements reacting to their overall pattern. As a result the composite unfolds and develops. My fascination throughout my entire career has been with the unfolding, the emergence of patterns, or what you might call evolution in a rather narrow sense. Lately this has become in turn an obsession with the evolution of technology.

For me all this happened in the early eighties. Something was in the air and I think for me this was just pure instinct, and I would say it goes an awful lot deeper. This is a matter of temperament or personality. Some people like to freeze what they are looking at. It's as if you can fast-freeze. You might be interested in how butterflies fly, and so you catch them and chloroform them, nail them to a board, and stare at their wings. But I was always interested in what makes the dynamics actually work.

My PhD was in operations research, essentially applied mathematics, in control theory. But how do you control systems that are unfolding over time? Most of my training has been in either engineering or mathematics, but I became fascinated with the economy. I don't know what was in the air. Sometime around 1980, we all got computers. By the mid-eighties these were what you would call workstations. I had a NeXT computer as soon as those things came out. Stuart Kauffman had a Sun SPARCstation. I remember a SPARC 2. But we all had workstations. You could simulate actual systems unfolding, so you could watch things unfold on your computer. You could write programs for each of these elements. They might be B cells or T cells in an immune system. They could be

W. Brian Arthur

species in Stuart's world or agents in the economy, investors and so on.

We began to extract parts of the world that we were interested in, re-create them on our computers, and then hit the return button for them to evolve. We watched what happened. We tried to get a mathematical description, because it wasn't sufficient to just say, well, here is how it unfolds on the computer. Here are nice pictures and we can freeze it here and we can look at these simulations or evolutions happening. It was better to see if we could get a mathematical version. Usually it was stochastic, probabilistic, and could go in different directions.

But there is some deeper part of my personality that became completely fascinated with the idea that the world is always unfolding. If somebody said to me, "Here is a picture of the universe as it is. Isn't it amazing? Isn't it incredible? Look at all the incredible structures that are there and they are frozen in time in this picture," I would say, "Yeah, it's interesting." But if somebody showed me how something was actually changing and unfolding, how new structures arise and fall away and further ones arise, I found that fascinating. I never knew how deep that went in my personality.

To give some personal history on where all this fits in: I was born in Northern Ireland, which might seem like a kind of fractured culture, and it was, but it was also a very stable culture. People knew who they were. The generation I'm from in Northern Ireland was known for writing and poetry. All the top poets were from the north of Ireland, not the Republic. I grew up in a very stable atmosphere but it was like a pressure cooker. It was fundamentalist Catholic. On the other side it was the same milieu, we had fundamentalist Protestantism. Something inside me just couldn't stand that. It was as if there were a lid on. When I was twelve or fifteen or twenty, before I left, I felt I could just simply scream. I felt that everything about the religious tensions in North Ireland hadn't changed in hundreds of years. Something was screaming at me. To cut a long story short, if that was a highly, highly ordered system in complexity terms, then there was the chaotic system.

I went to university in Belfast and got a superb education in engineering and then went to Berkeley, where I was essentially in applied mathematics and then economics. But it was as if I had gone from an ordered state into pure chaos in Berkeley. I couldn't handle that, either. That was far too disordered for me. The element I want to stress, and this is where I talk about people's personality, is that until you get older, into your thirties and forties and mature a bit—until I had a family and was teaching at Stanford—you really don't know what your temperament is all about. I knew I liked certain types of science, and I knew there were certain areas that appealed to me. But, as I said, a lot of this had to do with how populations grew and unfolded, human populations, how economies developed, and more lately how technology develops and evolves.

I must have gone through a midlife crisis sometime in my late thirties or early forties. I got an endowed chair at Stanford when I was thirty-seven. I had strived all my life to get to the next level and then the next level and then the next level and suddenly there weren't any more levels. I didn't quite know where to find my feet. I couldn't find anything I wanted to grab onto in Berkeley. The counterculture didn't appeal to me. This was still not long after the Vietnam War. I had worked in Austria. But there wasn't anything that I felt really formed a basis. I got rid of being Catholic, so what could I put my feet on?

I discovered something that was very much my temperament which surprised me. I got quite deeply involved in Taoism. The amusing thing about all of this is when I first started to get deeply into Taoism, it was at first a philosophy, and now it has mostly been more training, some of it martial arts, the Qi Gong, very much in the context of some of those Chinese movies you see: You wait at the door for three years, maybe I let you in, maybe teach you something, maybe not—this kind of thing. But when I got deeply into the philosophy, I began to realize it was all about things unfolding and things becoming. The deepest part of Taoist philosophy is saying there isn't really anything stable that you can grab onto. The whole world is always changing. The best thing you can do with your life is to go with whatever is happening, allow yourself to flow with it. I began to realize that all my research had the same theme. You could say that I found in a deep philosophy, a several-thousand-year-old philosophy, the counterpart of what interested me in science. Or maybe that's the way I thought all along, that this was deeply inside of me, this philosophy. Then I discovered its counterpart in science.

So I have been interested all my life in the development and unfolding of systems, not so much systems that are fixed in structure and behave mechanically like a car, or that are moving through time but not changing in structure. The way I was trained in economics, the structure of an economy, industries that are already out there, and the various parties involved—banking systems, regulatory systems, government, consumers—all of those structures are first taken for granted, then economists look at how that unfolds. That wasn't enough for me. I'm interested in how structures themselves change. In fact, economics didn't even say that much. Economics was essentially saying, given the fixed structure, where is its natural place to arrive at equilibrium? That was anathema from my point of view.

The Economics of Increasing Returns

Sometime in the late 1970s I started to read about the dynamics of enzyme chemistry, kind of an obscure thing. Two books I read in 1979 had a huge effect on me. One was *The Eighth Day of Creation* by Horace Freeland Judson—one of the best science books I've ever read in my life. The other one was Richard Rhodes's *The Making of the Atomic Bomb*. What Judson started to show me was that my view of biology was ignorant. I thought it was just about classification. Like stamp-collecting. Here is this species and that species and here's how this works and interacts with that. Judson was describing how molecular biology really worked and how all of that had been founded.

I had previously read James Watson's book on molecular biology. This brought the whole subject alive and made it all dynamic. It was all about interpreting the genetic code and the structure of hemoglobin. It was real science and I got tremendously excited by it. That turned me on to biology. Then I read Jacques Monod's book *Chance and Necessity* and began to realize that what he was describing were chemical reactions that were poised to go one way or another. These were autocatalytic reactions. There might be two end products possible in one of those reactions, and if the end product was A and A catalyzed more A or B catalyzed more B, then depending on which got ahead first, the whole thing could fall into more and more A catalyzing A, and B would be shut out. Or more and more B, shutting A out.

I also read some essays by Ilya Prigogine at the time. He was controversial because there was a lot of self-promotion. He was kind to me. In 1980, I made a pilgrimage to see Prigogine, and he took me out to the Royal Academy in Brussels: "As I was saying to the Pope on Thursday," and so on. Anyway, he was very kind to me. I find some people can have good ideas and very deep ideas but they maybe don't dot the i's and cross the t's quite as well as less-gifted people. I began to realize that what Prigogine and his group and Monod and François Jacob and others were talking about in France and Belgium were self-reinforcing systems—not ones that would run away like a snowball down a hill but ones that were more like bandwagons, any of which could gain momentum if they got off to a good start.

When you see a pattern like that, you begin to see it everywhere. The English language was like that. In the 1700s, if you wanted to make a bet, you would say, well, if we are sufficiently connected as a world, maybe a world language will appear. That had happened in the 1500s and 1600s. The language was Latin. But if you made your bet in the 1700s, you would have bet on French. The Russian court was speaking French, and French was fashionable in high circles. In the last half of the 19th century, you might have bet on German, because of the Austro-Hungarian Empire and the fantastic intellectual culture that, east of France, was sweeping right across Europe. Then English emerged in the 20th century. It has really taken over in this new century. It's the same thing. There is a lot of autocatalytic feedback, and I'm sure the rise of America did a lot to bring English in.

I began to remember that there was a huge set of questions that I had asked myself in graduate school while I was studying economics—not for a PhD but more as a hobby. It was my doctoral minor, and I had been taught very, very well by theorists in Berkeley—equilibrium theorists mostly. Economics basically says that everything arrives at an equilibrium, providing there are diminishing returns on the margin. So if something gets harder and harder to do—like you have to mine deeper and deeper for copper and it becomes more expensive to find because all the easy seams have been mined out—then you might start to substitute nickel for copper, and the economy reaches an equilibrium. If your time is competing for television and movies, once you run into diminishing returns in movies—you've seen all the good ones that are on—you tend to say, okay, I will balance that out with whatever, reading books or television or something. So providing we run into diminishing returns everywhere, the economy is assured of reaching equilibrium.

I had asked myself a question as a graduate student: What if there weren't diminishing returns? What if there were increasing returns? If something gets better the more you make it? What would happen? I thought of an example, pretty trivial. I had been to Hawaii as a graduate student. Milton Friedman was fashionable at the time, and I thought to myself: Suppose cars were offloaded for the first time from a ship in 1925, or some time like that, in Kauai, the most remote island. (I was in Kauai in the seventies when they installed their first traffic light.) Imagine there were a few dirt roads and Kauai got cars for the first time, and let's say the steering wheel in this thought experiment was in the middle of the car and there was no bias to the right or left. The car had just arrived and, say, Milton Friedman was running the island, so everybody was free to choose which side of the road they would drive on. It was a perfect libertarian island with no laws whatever. I thought, What would happen if I came back six months later?

I remember talking to Stuart Kauffman about this, and he and I agreed that what you would observe would be an awful lot of wreckage at the side of the road, but that cars would line up on one side or the other by a convention that eventually would emerge. Once more people started to drive, say, on the left, you would simply be a fool if you didn't do the same, because most of the cars you would meet if you drove on the right would be going the other way and you wouldn't survive that long. A system like that could tip one way or the other.

The examples I referred to were the QWERTY keyboard design and other structures in the economy that locked in more or less by chance. Economists were aware that increasing returns or positive feedback—positive feedback in this case being that the more cars drive on one side of the road, the more you are inclined to do the same. So it's an autocatalytic system. The more one candidate in a primary gets ahead, the more advantages he has, and so on. Systems like that tend to lock in to the dominance of one player or one side of the road or one typewriter keyboard, and everything else gets pushed aside. The lock-in happens probably by chance, by small random events getting magnified by these positive feedbacks.

At the time, economics was aware of the problem. The great economist Alfred Marshall in 1890 actually had written about it in a footnote, but he had asked: What if there were N firms, all with diminishing costs? They're all in the same industry. Aneroid barometers fascinated him—the more barometers a company got out, the cheaper it was for them to produce the next barometer. He posited that in a case of increasing returns or diminishing costs, after some time the market would have become dominated by one firm, which in his words would be whichever firm first got off to a good start. Economics couldn't say anything beyond that.

What I contributed was to show that, actually, we can say something beyond that. I realized that you could treat such dynamics as probabilistic, or technically as nonlinear stochastic processes, and look at how these by chance ended up with one outcome, or by chance with another. I started to work with Russian probability theorists. I didn't know enough sophisticated probability theory in the mid-eighties to solve these questions. But one colleague of mine from the famous Kolmogorov School in the then–Soviet

Union knew about this sort of probabilistic mathematics. I had to get up to speed myself. It took me a year or two. We cracked these problems. The first paper, on increasing returns, appeared in a Russian journal in 1983. All I could do was read my own name. I couldn't read anything else.

I had been told that all this stuff on increasing returns was "theoretical." Then I was talking about it in Santa Fe to some students and was walking to give my lecture and I had a complete epiphany. I thought, This isn't esoteric stuff, angels on pins. This applies to all of high-tech. I began to realize that all of high-tech operated according to increasing returns, which meant actually that the more a firm like Microsoft got ahead of the market, the more its brand would be out there, the more money it would have to parlay into the next thing. The more Google gets prominent as a search engine and the more people get used to using it, the more omnipresent Google becomes. Other search engines get pushed aside, like Ask.com or AltaVista. But you can't predict in advance which one it might be.

Suddenly I realized that there was a dynamic with Silicon Valley, and with high-tech all over the U.S., in which markets tended to tilt or tip to the dominance of one player. High-tech has what came to be known as "network effects": The more people that use Google, the more likely they are to use Google. This has huge upfront effects. Take Microsoft. They used to give you Windows on one of these little disk things, but the first copy of Windows NT, or Windows 2000, or whatever it would be these days might cost Microsoft something like $2 billion. The next copy might cost them fractions of a cent. So their costs very rapidly go down per unit the more they get out there. But this is not true for dog food. The next unit of dog food costs just about as much per unit as the first one.

W. Brian Arthur

Ideas operate very much according to increasing returns. It's costly for someone to dream them up and almost free for anyone to distribute them. It turns out that information is virtually free, and we are seeing, as the economy runs more and more on ideas rather than on bulk commodities such as processed corn, processed iron ore, steel, or cars, that we have different rules. Increasing rather than diminishing returns. A different economics applies.

We are coming to a commerce of ideas, i.e., an interchange of ideas that are simply out there. People will use them as a kind of coinage. We will exchange ideas. We can use other people's ideas to construct yet further ideas. There is a natural tendency for information to be free. I don't know if it wants to be anything, but certainly once it's out there it's fairly costless to swap around. There are some superficialities, and we are going to learn that things are much more complicated than that. Information is free, but grokking information, being able to take it in and use it for yourself, is not free. It's not sufficient to tell somebody something. That's just an idea. Ideas have to come with backing, and they have to come with understanding. It's a bit like if I showed you a Chinese painting. You, or somebody who is skilled, could reproduce it, but what they couldn't reproduce is the depth of understanding and the culture, the context, the history, everything that led up to it.

A lot of complexity has to do with how systems unfold and by chance could go one way or the other. I was asking the same questions in economics, given positive as well as negative feedbacks: How do certain patterns form in the economy? How do we end up driving on the right rather than on the left? How come we speak English? But more to the point, how does Silicon Valley end up in a bunch of apricot orchards rather than somewhere over in the East Bay? How do we get the dominance of something like

Microsoft? Those were questions of unfolding, but they weren't trivial questions such as saying there is another stage and another stage. These were systems that according to chance could fall into certain patterns and then the next pattern could fall on top of that.

There is another element to high-tech. Microsoft may have been the thing of the moment in the early nineties, locking in most of the tech market with huge, increasing returns. But just when you think it is going to be forever, something else comes along, Google. What fascinated me was that the economy had kind of layer after layer after layer, like an archaeological site, Troy, with fourteen layers of cities. But you could watch them in your own lifetime, five or six of these layers forming on top of each other.

I became fascinated by another question. We take an awful lot of things in the world for granted. We take it for granted that as technologies progress, they become more complicated. Something told me in 1993 to read up about jet engines. I thought, I'm not interested in jet engines, but it was an instinct. Of course, within two days I began to realize that the original jet engine was quite simple and the ones we have now are tens of thousands of parts and really, really complicated. You could say, oh, that's just the way technology is, but I began to wonder why it was that way.

By the mid-nineties, a series of questions emerged that I really started to obsess about. What, really, is technology? I had no clue. Somehow I began to realize that the economy forms out of its technologies in some way. The difference between a high-tech economy, which we have here, and, say, the one in the Trobriand Islands, is the difference between having sophisticated technology vs. fishing and canoes. I wondered how technologies developed over time—like how the jet engine became more sophisticated and why that should be. It's like saying, Why is the sky blue? We just

W. Brian Arthur

take it for granted. Then above all I began to wonder, Was there a theory of evolution for technology?

I was very aware as a graduate student that the economy somehow arose from its technologies. That wasn't new. Karl Marx had written a lot about that, as had other really good economists in the mid-1800s. But we tend to see an economy as out there and fixed: It uses factories and factories use technologies and machines and things like that. I began to wonder, How does something we call the economy arise out of technologies? I began to realize that every technology I knew about, from the computer to jet engines, started off pretty simple and wound up incredibly complicated, orders of magnitude more complicated than what it started off as. Then there was a question that had been hanging in the air since the 1860s, posed by archaeologists and anthropologists and a guy called Augustus Pitt Rivers, and by Samuel Butler: Does technology as a whole evolve?

Darwin had answered that question in biology. The many species, in what we now call the biosphere, are somehow all related by threads of common ancestry that go way back. That wasn't new to Darwin. His grandfather had said as much, and other people had said as much in the 1700s and early 1800s. What Darwin was able to supply was the mechanism by which that branching and then speciation had occurred. His book was about the origin of species. It wasn't called "Evolution."

Take various technologies: jet engines, the refining of certain types of metals, the steam engine, lasers, computers, or methods for doing things, like sorting algorithms. If you take all of those technologies together, can there be a theory of evolution by which they are somehow related, by threads of common ancestry, to earlier technologies?

Economists, historians, and technology thinkers have for decades been trying to apply Darwin's theory to technology, saying, Okay, different designers have different ways to solve problems. That gives you variation. Then the better solutions are selected. So you have variation and selection in technology. That would give you a theory of evolution for technology. I'm condensing 150 years worth of thinking since Darwin's book.

But those theories weren't satisfactory. At least they didn't satisfy me. The jet engine didn't evolve out of variations of the air piston engine. In Darwin's scheme, if you get a new species of finch, it's related to some older species of finch, but it has adapted gradually, the structure of its beak changing because of some circumstance. There is a gradual progression until a new species is formed in a slightly different niche and branches off from the old species. You couldn't say that the laser or the jet engine branched off from anything. They were completely new. So I began to realize that there wasn't a satisfactory theory of evolution for technology. In fact, there wasn't a satisfactory theory of anything I could think of in technology.

There is no theory of technology. People greeted the very idea that there could be a theory of technology either with, "I guess you could do that, but why would you want to?" Or, "I don't know, do we need anything like that? Do we need a theory of technology?" I didn't start to do this academically. It was kind of an obsession. I began to think, yeah, there are some common principles that I could use to think about all of this. So I went underground. I went back to my lab notebooks a few weeks ago when I finished the project, because I was curious to see how long I had been working on this. When I looked at my lab books, I realized I had been working on it for thirteen years. But most of the work was certainly not writing. It was reading and thinking about technology and the history of how technologies came into being—very specific ones, maybe a dozen to twenty technologies, everything from computers to search algorithms. Certain technologies I learned in great detail, just as a biologist would have to study certain types of beetles or something to make sense of his or her arguments.

What happened was that I went underground. I once read that the mathematician Andrew Wiles, who proved Fermat's Last Theorem, did exactly that. Wiles, I think out of instinct, as a very good Princeton mathematician, decided that he didn't want his embryonic thoughts to be hammered by criticism. He needed space and time to think out his ideas; he needed to put things together and not be bothered and questioned all the time. He didn't want anyone to know what he was working on, because it might lead to a frenzy of competition. For ten or more years, as far as I know, he worked in the attic of his house in Princeton with one or two colleagues who were in on the plot and bit by bit built up his understanding until he could construct a working version of his theorem. Then he would give it out to a wider set of people to probe and test and so on. I made no such resolution when I started to work on my own project around 1996. But it is essentially what happened. I told very few people. The word came back from the Santa Fe Institute—what has Arthur really been doing? He has some very interesting early work and some promising work, but we haven't heard anything from him in years.

At any rate, for better or worse, I became convinced that there were many questions unanswered in technology. What, really, is technology? How does it work? How does it develop? Is there an overall theory of its evolution? Technology was actually a gold mine.

I began to realize that all new species of technologies, such as the laser printer that evolved out of Xerox PARC, are composed of already existing elements. When Gary Starkweather started to work on the laser printer down in Palo Alto, he was basically saying, we don't want line printers. They can't print images. They can't change font size. They are basically computer-driven type-writers. After a lot of thinking about other ideas—say, writing on cathode-ray screens and all—he had the idea that maybe he could get a computer to control a very highly focusable laser beam and write (or the word they would have used is paint) an image onto a Xerox drum. So the elements existed: a computer, computer-controlled lasers, the elements to control the laser operation, and xerography. And we got the laser printer.

When I started to look at any new species of technology—be it a jet engine or laser printer or sorting algorithm—I found all their components already existed (or could be constructed from things that existed). I began to realize that it was possible to put a theory of evolution together with combination at the heart of it. So what I'm saying is that technologies evolve from previous technologies by selecting building blocks and putting them together in new ways.

Some people had realized this, and some had written a bit about it, but most of the writing was maybe a paragraph here and maybe a few sentences there, by somebody really smart, like Schumpeter, who, working a hundred years ago, never went beyond some preliminary ideas that are now very widely quoted. Nobody had worked out this idea about combination. I came to it independently, but didn't do much about it in the eighties. Then I discovered a lot of kind of hand-wavy literature in which people talked about it.

I was faced with another question: If everything is a combination of something that existed before, why aren't sophisticated technologies like magnetic resonance imaging constructed out of flint or obsidian or whatever we had 10,000 years ago? It's not the original components of 10,000 years ago that are getting combined to give us MRIs. I began to realize there is another leg to it—that every so often technologies are used to capture phenomena, in this particular instance nuclear magnetic resonance.

Every so often we use instruments and modalities, methods and technologies, to capture some phenomenon—say, the ability to affect the nuclear spin of certain atoms and use that to make certain measurements. It becomes some sort of diagnostic technique. So the two legs of the theory of evolution that are in technology are not at all Darwinian. They are quite different. They are (1) that certain existing building blocks are combined and recombined, and (2) that every so often some of those technologies get used to capture novel, newly discovered phenomena, which are in turn encapsulated as further building blocks. Most new technologies that come into being are only useful for their own purpose and don't form other building blocks. But occasionally, some do.

What really excites me about this is not so much technology. What really interests me is that astrophysicists and cosmologists have a very similar idea, both of the formation of life on earth and of the formation of the universe. Now this is off my beat, so maybe I'm a little hazy on this, but after the Big Bang, if you wait long enough—maybe 10-to-the-minus–27 seconds, eternities of very short magnitude—elementary particles, quarks, begin to combine to form the basic hadron building blocks, which further combine to give you atoms and the molecules, which over time lead to very rapidly expanding gases, which in turn form stellar systems. All of these steps are formed by combinations of elements forming new building blocks that give you further combinations.

The same is true of early life. I have been talking to people recently at the Santa Fe Institute, and they are talking about various reactions in what could be called the evolution of metabolic pathways. You get terribly simple metabolisms forming very early life, like 4 billion years ago. Then they form certain elements that in combination can give you further elements that are catalyzed by some building blocks that give you yet newer molecules. So the whole of what we call life, building up to RNA and then DNA, forms out of structures that are combinations of simpler ones that give you combinations of yet further ones.

So in one phrase, to go back to technology, my argument up until my project, I think it's safe to say, was that if you asked people, "What is technology?" on the whole they would have said it's a bunch of standalone methods or devices—the Solvay process, the computer, laser printers, and so on—that are sometimes interrelated and have some sort of ancestry.

What I'm saying is, no, all of this forms a gigantic chemistry. The simpler molecules that have formed in technology—the computer, the laser, xerography—can be put together to form yet a new molecule, the laser printer, which can be put together to form something more complicated. Technology viewed as a whole is chemistry, and its chemistry is still building.

15.

Aristotle

The Knowledge Web

W. Daniel Hillis

Physicist; computer scientist; chairman, Applied Minds, Inc.; author,
The Pattern on the Stone

I have always envied Alexander the Great, because he had Aristotle as a personal tutor. In those days, Aristotle knew pretty much everything there was to know. Even better, Aristotle understood the mind of Alexander. He understood which topics interested Alexander, what Alexander knew and did not know, and what kinds of explanations Alexander preferred. Aristotle had been a student of Plato, and he was himself a great teacher. We know from his writings that he was full of examples, explanations, arguments, and stories. Through Aristotle, Alexander had the knowledge of the world at his command.

Of course, no one today knows all that is known, in the sense that Aristotle did. Now there is far too much knowledge for that to be possible. The scientific revolution, and the technological revolution that followed it, led to a self-reinforcing explosion of knowledge. The explosion continues. Today not even the most highly trained scientist, the most scholarly historian, or the most competent engineer can hope to have more than a general overview of what is known. Only specialists understand most of the new discoveries in science, and even the specialists have trouble keeping up.

This problem isn't new. In 1945, Vannevar Bush wrote an essay for *The Atlantic Monthly* about the problem of too much knowledge. He wrote:

> There is a growing mountain of research. But there is increased
> evidence that we are being bogged down today as specialization
> extends. The investigator is staggered by the findings and
> conclusions of thousands of other workers—conclusions that he
> cannot find time to grasp, much less to remember, as they appear. Yet
> specialization becomes increasingly necessary for progress, and the
> effort to bridge between disciplines is correspondingly superficial.

Bush's imagined solution to this problem was something he called Memex. Memex was envisioned as a system for manipulating and annotating microfilm (computers were just then being invented). The system would contain a vast library of scholarly text that could be indexed by associations and personalized to the user. Although Memex was never built, the World Wide Web, which burst onto the scene half a century later, is a rough approximation of it.

As useful as the Web is, it still falls far short of Alexander's tutor or even Bush's Memex. For one thing, the Web knows very little about you (except maybe your credit card number). It has no model of how you learn, or what you do and do not know—or, for that matter, what it does and does not know. The information in the Web is disorganized, inconsistent, and often incorrect. Yet for all its faults, the Web is good enough to give us a hint of what is possible. Its most important attribute is that it is accessible, not only to those who would like to refer to it but also to those who would like to extend it. As any member of the computer generation will explain to you, it is changing the way we learn.

W. Daniel Hillis

A New Tool for Learning

Let's put aside the World Wide Web for a moment to consider what kind of automated tutor could be created using today's best technology. First, imagine that this tutor program can get to know you over a long period of time. Like a good teacher, it knows what you already understand and what you are ready to learn. It also knows what types of explanations are most meaningful to you. It knows your learning style: whether you prefer pictures or stories, examples or abstractions. Imagine that this tutor has access to a database containing all the world's knowledge. This database is organized according to concepts and ways of understanding them. It contains specific knowledge about how the concepts relate, who believes them and why, and what they are useful for. I will call this database the knowledge web, to distinguish it from the database of linked documents that is the World Wide Web.

For example, one topic in the knowledge web might be Kepler's third law (that the square of a planet's orbital period is proportional to the cube of its distance from the sun). This concept would be connected to examples and demonstrations of the law, experiments showing that it is true, graphical and mathematical descriptions, stories about the history of its discovery, and explanations of the law in terms of other concepts. For instance, there might be a mathematical explanation of the law in terms of angular momentum, using calculus. Such an explanation might be perfect for a calculus-loving student who is familiar with angular momentum. Another student might prefer a picture or an interactive simulation. The database would contain information, presumably learned from experience, about which explanations would work well for which student. It would contain representations of many successful paths to understanding Kepler's law.

Given such a database, it is well within the range of current technology to write a program that acts as a tutor by selecting and presenting the appropriate explanations from the database. The automated tutor would not need to create the explanations themselves—human teachers would create the explanations in the knowledge web, as well as the paths that connect them. The program would merely find the appropriate paths between what a student already knows and what he or she needs to learn. Along the way, the automatic tutor would quiz the student and respond to questions, much as a human tutor does. In the process, it would improve both its model of the student and the information in the database about the success of the explanations.

For example, imagine yourself in the position of an engineer who is designing a critical component and wants to learn something about fault-tolerant design. This is a fairly specialized topic, and most engineers are not familiar with it; a standard engineering education treats the topic superficially, if at all. Fault-tolerant design is an area normally left to specialists. Unless you happen to have taken a specialized course, you are faced with a few unsatisfactory alternatives. You can call in a specialist as a consultant, but if you don't know much about the field it's difficult to know what kind of specialist you need, or if the time and expense are worth the trouble. You could try reading a textbook on fault-tolerant design, but such a text would probably assume a knowledge you may have forgotten or may never have known. Besides, a textbook is likely to be out of date, so you will also have to find the relevant journals to read about recent developments. If you find them, they will almost certainly be written for specialists and will be difficult for you to read and understand. Given these unsatisfactory choices, you will probably just give up. You will go ahead and design the module without benefit of the proper knowledge, and hope for the best.

W. Daniel Hillis

Learning with *Aristotle*

Now let's assume that you have access to the automatic tutor. Let's call it *Aristotle*.

Aristotle would begin by asking you how much time you're willing to devote to this project and the level of detail you want. Then *Aristotle* would show you a map of what you need to learn. The tutor program does this by comparing what you know to what needs to be known to design fault-tolerant modules. It knows what needs to be known because this is a common problem faced by many engineers, and knowledgeable teachers have identified the key concepts many times. *Aristotle* knows what you know because it has worked with you for a long time. There may be some things you're familiar with that *Aristotle* doesn't know you know, but you can point these things out to *Aristotle* when it shows you the learning plan. *Aristotle* might take your word for what you know, but it is more likely to quiz you about some of the key concepts, just to make sure.

Aristotle plans its lessons by finding chains of explanations that connect the concepts you need to learn to what you already know. It chooses the explanatory paths that match your favorite style of learning, including enough side paths, interesting examples, and related curiosities to match your level of interest. Whenever possible, *Aristotle* follows the paths laid down by great teachers in the knowledge web. *Aristotle* probably also has a model of how you want to be paced: when you have learned enough for one day, when it needs to throw in an interesting side story, etc. Along the way, *Aristotle* will not only explain things to you but will also ask you questions—both to make you think and to verify for itself that concepts are being learned successfully. When an explanation doesn't work, *Aristotle* tries another approach, and of course you can always ask questions, request examples, and give *Aristotle* explicit feedback on how it's doing. *Aristotle* then uses all these forms of feedback to adjust the lesson, and in the process it learns more about you.

The process of teaching helps *Aristotle* learn to be a better teacher. If an explanation doesn't work and consistently raises a particular type of question, then *Aristotle* records this information in the knowledge web, where it can be used in planning the paths of other students. The feedback eventually makes its way back to the knowledge web's human authors, so that they can use it to improve their explanations.

Like any good tutor, *Aristotle* allows you to get sidetracked from a lesson plan and follow your interests. If you find a particular example compelling, you may want to know more about it. If some concept you have just learned allows you to appreciate an elegant explanation of some fact you already know, *Aristotle* may point it out. If you are close to understanding something of critical importance or something that would be of particular interest to you, *Aristotle* may decide to show it to you even though it is not strictly necessary as part of the lesson. Of course, as *Aristotle* gets to know you, it will know how much you like this sort of distraction.

Once you have learned the material, and *Aristotle* has verified that you have learned it, the program will update its database to indicate that you have recently learned it. As you learn more and more, it will continue to connect your recently acquired knowledge to the new concepts you are learning, until you have fully integrated them. Because *Aristotle* knows which subjects you are and have been interested in, it can consolidate your learning by finding connections that tie these subjects together.

For example, there's a short film of Dr. Richard Feynman explaining a principle of quantum mechanics called Bell's inequality. Most people have little interest in quantum mechanics and no interest at all in understanding Bell's inequality. Most quantum physicists already understand Bell's inequality and would learn little from Feynman's explanation. On the other hand, if you are a student who is just learning quantum mechanics, who has just mastered the necessary prerequisites, Feynman's explanation can be exciting, startling, and enlightening. It not only can explain something new but can also help you make sense of what you have recently learned. The trick is showing the film clip at just the right time. *Aristotle* can do that.

I used an engineering example to describe *Aristotle* because most engineering knowledge is a straightforward, factual type of knowledge. Similar techniques would work for learning other subjects—history, mathematics, or the kind of technical information that might normally be conveyed in a training course or a technical manual. Of course, there are types of useful knowledge that a program like *Aristotle* would not be suited to: *Aristotle* would not be of much use in learning how to ride a bicycle or tell a joke. It would not replace hands-on experience, nor would it replace the enthusiasm and wisdom of a great teacher. What *Aristotle* would do is help you gain mastery of factual knowledge—exactly the kind of knowledge that is overwhelming us.

How the Knowledge Web Changes Education
In his book *The Diamond Age*, the science-fiction writer Neal Stephenson describes an automatic tutor called the Primer that grows up with a child. Stephenson's Primer does everything described above and more. It becomes a friend and playmate to the heroine of the novel, and guides not only her intellectual but also her emotional development. Such a Primer is beyond the capabilities of current technology, but even a program as limited as *Aristotle*

would be a step in that direction. Teachers and students both understand that a school is no longer able to "preload" its students with the knowledge that they will need for life. Instead, a good school teaches the basics—reading, arithmetic, social skills—and introduces students to subjects that they can learn more about. It gives them an overview of knowledge as a starting point for further learning. A good school education also gives students the skills to acquire knowledge as they need it.

Teachers know that individual attention helps a child learn, and they would like to give their students more of it. Even a young child has special interests, special topics that he or she would like to know more about. A good teacher learns to recognize these individual interests and tries to nurture them, but this takes a lot of time. A program like *Aristotle* will give teachers a tool to help children follow their passions. It will also enable teachers to evaluate a child's progress and to provide individualized instruction in areas in which the child has gaps. A computer program like *Aristotle* cannot replace most of what goes on in school, but it can complement what goes on there. It can free teachers from the routine job of broadcasting information and give them more time to provide individual attention to their students.

A system like *Aristotle* also empowers teachers by giving them a way to publish. Any good teacher knows how to teach certain topics especially well, but there are few easy ways for them to share that information effectively with others. A teacher can write a textbook, or develop a curriculum, but each of those efforts is a major undertaking. There is no simple way for a teacher to publish an isolated idea about how to explain something. If a system like *Aristotle* existed, then with the proper authoring tools a teacher could publish a single explanation—an effort comparable to creating a Web page. In fact, existing Web pages are a good source of initial content for the knowledge web. As Marshall McLuhan said, "The content of the new medium is the old medium." The initial content of the knowledge web will be the old curriculum materials, textbooks, and explanatory pages that are already on the World Wide Web. The existing materials already contain most of the examples, problems, illustrations, and lesson plans that the knowledge web will need.

As students gain access to the best explanations from the best teachers on a given subject, their own teachers will be able to take on the role of coaches and mentors. Freed from the burden of presenting the same information over and over, the teachers will be able to give greater individual attention to their students.

A Better Infrastructure for Publishing

The shared knowledge web will be a collaborative creation in much the same sense as the World Wide Web, but it can include the mechanisms for credit assignment, usage tracking, and annotation that the Web lacks. For instance, the knowledge web will be able to give teachers and authors recognition or even compensation for the use of their materials. Teachers and learners will be able to add annotations and links to explanations—connecting them to other content, suggesting improvements, or rating their accuracy, usefulness, and appropriateness for children of various ages. For instance, with such a system it would be possible for the learner to accept only knowledge that had been certified as correct by an authority such as *Encyclopædia Britannica* or the National Academy of Sciences.

All of this raises the possibility of a different kind of economic underpinning for the knowledge web, one that is not possible on the document Web of today. The support infrastructure for payments would allow different parts of the knowledge web to operate in different ways. For instance, public funding might pay for the creation of curriculum materials for elementary school teachers and students, but specialized technical training could be offered on a fee or subscription basis. Companies could pay for encoding the knowledge necessary to train their employees and customers, consultants would be able to publish explanations as advertising for their services, and enthusiasts would offer their wisdom for free. Students could subscribe not only to particular areas of knowledge but to particular types of annotations, such as commentary or seals of approval. Schools and universities could charge for interaction with teachers and certification of the student's knowledge. The system could also become the ultimate hiring tool, since employers could map areas of knowledge that they need in prospective employees.

What Makes the Knowledge Web Different?
One way to think about the knowledge web is to compare it with other publishing systems that support teaching and learning. These include the World Wide Web, Internet news groups, traditional textbooks, and refereed journals. The knowledge web takes important ideas from these systems. These ideas include peer-to-peer publishing, vetting and peer review, linking and annotation, mechanisms for paying authors, and guided learning. Each of these existing media demonstrates the success of one or more of these ideas. They are all incorporated into the knowledge web.

Peer-to-Peer Teaching
One of the reasons why Internet news groups and the World Wide Web have enjoyed such runaway success is that they allow people to communicate with each other directly, without publishers as

intermediaries. The great advantage of such a peer-to-peer publishing system is that anyone with something interesting to say has an easy way to say it. The Internet has eliminated the publishing bottleneck and has created a flood of authorship. This basic human desire to share knowledge is what will drive the creation of the knowledge web. The task of recording the world's knowledge is so overwhelming that only peer-to-peer publishing can plausibly accomplish it. Yet the knowledge web is not only a record of knowledge but also a way of imparting it. The knowledge web will do for teaching what the World Wide Web did for publication. Peer-to-peer teaching will allow literally millions of people to help each other learn.

Vetting and Peer Review

One of the downsides of peer-to-peer publishing is quality control. Publishers of textbooks and journals do more than market and distribute; they also edit and select. In the case of peer-reviewed journals, some of the burden of quality control is shifted to the reviewers, but it is still coordinated by the publisher. On the World Wide Web, there is no commonly accepted system of rating and peer review, nor is there a mechanism to support one. The result is chaos. The information you find by searching on the World Wide Web is often irrelevant, badly presented, or just plain wrong. It is difficult to screen out obscene material and propaganda. It is almost impossible to sort the wheat from the chaff.

The knowledge web addresses this problem by supporting an infrastructure for peer review and third-party certification. It supports mechanisms for the labeling, rating, and categorization of material, both by the author and by third parties. The browsing tools will allow information to be filtered, sorted, and labeled according to these annotations. In addition, user feedback tools will be built into the browsing software to help identify material that is particularly good, bad, or controversial.

Linking and Annotation

Anyone who has used the World Wide Web understands the importance of linking. In principle, articles in conventional journals also support a kind of linking, in the form of footnotes and references, but these kinds of links are far less convenient to use than the convenient "click throughs" of hypertext. Even the simple message-thread linking of threaded news groups helps make the information more usable. Students accustomed to hypertext find the linear arrangement of textbooks and articles confining and inconvenient. In this respect, the Web is clearly better. The knowledge web will allow an even more generalized form of linking than the World Wide Web. In the knowledge web, not only the author but also third parties can create links, comments, and annotations.

Ways to Pay Authors

An advantage of textbooks and journals over the World Wide Web is that they support a mechanism for paying the author. The World Wide Web has demonstrated that many authors are willing to publish information without payment, but it does not give them any convenient option to do otherwise. The knowledge web will provide that option by supporting various payment mechanisms, including subscription, pay per play, fee for certification, and usage-based royalties. It will also support and encourage the production of free content.

Much of the content on the knowledge web will probably be free, but there are a number of other economic models that can coexist with this. One of the most obvious is the paid course, in which a student pays tuition for a cluster of services, including access to teachers, curriculum materials, and interaction with other students, and some form of certification at the end. With the knowledge web, many institutions may choose to offer the curriculum material for free—as a form of advertising—and charge for the other services, especially the certification. The knowledge web would also help solve one of the online course providers' greatest problems, which is marketing. The knowledge web would help direct students to the courses that meet their needs.

Another model that may work well is a micropayment system, in which a student pays a fixed subscription fee for access to a wide range of information. Usage statistics would serve as a means to allocate the income among the various authors. This system has the advantage of rewarding authors for usefulness without penalizing students for use. The students' fees would be independent of the amount of use they make of the system. The ASCAP music royalty system and university payments for student access to *Encyclopædia Britannica* are examples of how such a system might work.

Guided Learning

Part of the information in a textbook is about the subject matter, but part of it is also about how to learn the subject matter. A good textbook is full of information on the plan of attack, strategies for learning, practice problems, and suggestions for further study. Part of what makes up a good curriculum is not just the material but the plan for moving through it. A textbook often has an accompanying "teacher guide" that contains more of this type of information. For example, the guide may note that if the student makes a certain type of error, the student is probably missing a particular point and needs it to be re-explained. Some of the best computer-aided instructional material also encodes such information. The knowledge web provides an easy mechanism for a teacher to include this type of information.

TABLE OF AFFORDANCES

	The Web	News Groups	Textbooks	Journals
Peer-to-Peer publishing	Yes	Yes	No	Limited
Supports linking	Yes	Limited	No	Limited
Ability to add annotations	No	Yes	No	No
Vetting and certification	No	Limited	Yes	Yes
Supports payment model	No	No	Yes	Yes
Supports guided learning	Limited	No	Yes	No

Achieving Critical Mass

Once the knowledge web achieves critical mass, it is easy to see how it can sustain itself. The real question is how it gets started. Presumably, the first users will be adults, and the first adopters will be industry and government. U.S. industry spends $60 billion a year training employees. It spends even more on customer support, product liability, poor design, and other costs that could be mitigated by better training. This is the first market for the knowledge web. The second market is probably the military, which is the single largest educator of adults. The third market is continuing education for individuals.

Adults are an easier initial market than children for several reasons. Adults are usually learning because they want to or because they need to. They are already motivated to learn. Adults often need specific knowledge for a specific reason. A recent survey of

W. Daniel Hillis

the American workplace showed that 80 percent of workers feel that additional education is important for them to be successful at their job. Adults are much more likely to treat time as a limited resource. They want to learn efficiently. Also, the mechanisms that pay for adult education are often more rational and less politically charged than the mechanisms that pay for the education of children. For-profit companies are likely to adapt quickly to a more efficient process.

Eventually traditional educational institutions will use this new system of learning. It will be used first by colleges and trade schools; then it will make its way into secondary and elementary education. It is important to emphasize that computers will not replace the teachers; rather, they will give teachers a new tool. Instead of spending most of their time broadcasting information to a group, teachers will have more time to help students integrate their knowledge through discussion and through individualized interaction.

There are three technical components necessary for such a system to work: the tutor (browser), the authoring tool, and the knowledge web itself. This last component is the most difficult to create, but fortunately it does not have to be built all at once. Even a small part of it would be useful. Presumably the knowledge web will get its start in a few narrow areas, probably determined by available funding. For instance, it is easy to imagine a scenario in which a part of the knowledge web is initially funded by a supplier to explain the use of its products. Imagine, for example, that Cisco publishes the knowledge of how to configure and maintain its routers in this form. Another scenario is that the government sponsors the creation of the system for a specific application, such as continuing education for schoolteachers or job training for factory workers. A company might pay for the development of training programs for its workers and customers. A foundation might sponsor an initial effort as a way to have an impact on education.

Summary: An Idea Whose Time Has Come

It seems almost inevitable that such a system will eventually be developed, but why is now the time to undertake such a project? After all, people have been dreaming of such a project for half a century, yet Memex, XanEdu, Plato, WAIS, and numerous other schemes have failed to achieve critical mass. Why should this one succeed? For one thing, the infrastructure is now ready for it. The knowledge web requires widespread access to network-connected computers capable of handling graphics, audio, and video. This did not exist until recently. But it is not just the technology that is ready for this idea—people are ready for it. E-mail, the Web, and video games have all whetted their appetites. The younger generation is more than ready. They expect something better than listening to lessons in class-sized groups.

At the same time that a solution is becoming possible, the problem is reaching a crisis point: The amount of knowledge is becoming overwhelming, and the need for it is increasing. There is a widespread conviction that something radical needs to be done about education—both the education of children and the continuing education of adults. The world is becoming so complicated that schools are no longer able to teach students what they need to know, but industry is not equipped to deal with the problem, either. Something needs to change.

With the knowledge web, humanity's accumulated store of information will become more accessible, more manageable, and more useful. Anyone who wants to learn will be able to find the best and the most meaningful explanations of what they want to know. Anyone with something to teach will have a way to reach those who want to learn. Teachers will move beyond their present role as dispensers of information and become guides, mentors, facilitators, and authors. The knowledge web will make us all smarter. The knowledge web is an idea whose time has come.

16.

The Pancake People

Or, "The Gods Are Pounding My Head"

Richard Foreman

vs. The Gödel-to-Google Net

George Dyson

Richard Foreman: *Founder and artistic director, Ontological-Hysteric Theater*

Foreman has written, directed, and designed more than fifty of his own plays, both in New York City and abroad. Five of his plays have received Obie Awards for best play of the year—and he has received five other Obies for directing and for "sustained achievement." In 2005, Forman directed the surrealist play The Gods Are Pounding My Head.

George Dyson: *Science historian; author,* Darwin Among the Machines *and* Project Orion

The Pancake People
Or, "The Gods Are Pounding My Head"

A Statement

When I began rehearsing *The Gods Are Pounding My Head*, I thought it would be totally metaphysical in its orientation. But as rehearsals continued, I found echoes of the real world of 2004 creeping into many of my directorial choices. So be it.

Nevertheless, this very—to my mind—elegiac play does delineate my own philosophical dilemma. I come from a tradition of Western culture in which the ideal (my ideal) was the complex, dense, and "cathedral-like" structure of the highly educated and articulate personality—a man or woman who carried inside him- or herself a personally constructed and unique version of the entire heritage of the West.

And such multifaceted, evolved personalities did not hesitate—especially during the final period of "Romanticism-Modernism"—to cut down, like lumberjacks, large forests of previous achievement in order to heroically stake a new claim to the ancient inherited land—this was the ploy of the avant-garde.

But today, I see within us all (myself included) the replacement of complex inner density with a new kind of self evolving under the pressure of information overload and the technology of the "instantly available." A new self that needs to contain less and less of an inner repertory of dense cultural inheritance—as we all become "pancake people"—spread wide and thin as we connect with that vast network of information accessed by the mere touch of a button.

Will this produce a new kind of enlightenment or "super-consciousness"? Sometimes I am seduced by those proclaiming so—and sometimes I shrink back in horror at a world that seems to have lost the thick and multitextured density of deeply evolved personality.

But, at the end, hope still springs eternal . . .

A Question

Can computers achieve everything the human mind can achieve?

Human beings make mistakes. In the arts—and in the sciences, I believe?—those mistakes can often open doors to new worlds, new discoveries and developments, the mistake, themselves becoming the basis of a whole new world of insights and procedures.

Can computers be programmed to "make mistakes" and turn those mistakes into new and heretofore unimaginable developments?

The Gödel-to-Google Net

Richard Foreman is right. Pancakes indeed!

He asks the Big Question, so I've enlisted some help: the Old Testament prophets Lewis Fry Richardson and Alan Turing; the New Testament prophets Larry Page and Sergey Brin.

Lewis Fry Richardson's answer to the question of creative thinking by machines is a circuit diagram, drawn in the late 1920s and published in 1930, illustrating a self-excited, nondeterministic circuit with two semi-stable states and captioned "Electrical Model Illustrating a Mind Having a Will but Capable of Only Two Ideas."

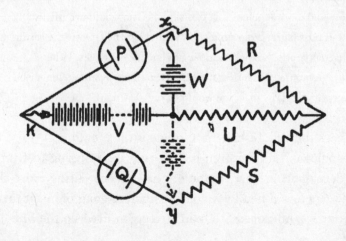

Machines that behave unpredictably tend to be viewed as malfunctioning, unless we are playing games of chance. Alan Turing, namesake of the infallible, deterministic, universal machine, recognized (in agreement with Richard Foreman) that true intelligence depends on being able to make mistakes. "If a machine is expected to be infallible, it cannot also be intelligent," he argued in 1947, drawing this conclusion as a direct consequence of Kurt Gödel's 1931 results.

"The argument from Gödel's [theorem] rests essentially on the condition that the machine must not make mistakes," he explained in 1948. "But this is not a requirement for intelligence." In 1949, while developing the Manchester Mark I for Ferranti Ltd., Turing included a random-number generator based on a source of electronic noise, so that the machine could not only compute answers, but occasionally take a wild guess. Turing observed:

> Intellectual activity consists mainly of various kinds of search. Instead of trying to produce a programme to simulate the adult mind, why not rather try to produce one which simulates the child's? Bit by bit one would be able to allow the machine to make more and more "choices" or "decisions." One would eventually find it possible to program it so as to make its behaviour the result of a comparatively small number of general principles. When these became sufficiently general, interference would no longer be necessary, and the machine would have "grown up."

That's the Old Testament. Google is the New.

Google (and its brethren metazoans) are bringing to fruition two developments that computers have been waiting more than sixty years for. When John von Neumann's gang of misfits at the Institute for Advanced Study in Princeton fired up the first 32 ×

32 × 40-bit matrix of random access memory, no one could have imagined that the original scheme for addressing these 40,960 ephemeral bits of information, conceived in the annex to Kurt Gödel's office, would now have expanded, essentially unchanged, to address all the information contained in all the computers in the world. The Internet is nothing more (and nothing less) than a set of protocols for extending the von Neumann address matrix across multiple host machines. Some 15 billion transistors are now produced every second, and more and more of them are being incorporated into devices with an IP address.

As all computer users know, this system for Gödel-numbering the digital universe is rigid in its bureaucracy, and every bit of information has to be stored (and found) in precisely the right place. It is a miracle (thanks to solid-state electronics, and error-correcting coding) that it works. Biological information processing, in contrast, is based on template-based addressing, and is consequently far more robust. The instructions say "Do X with the next copy of Y that comes around" without specifying which copy, or where. Google's success is a sign that template-based addressing is taking hold in the digital universe, and that processes transcending the von Neumann substrate are starting to grow. The correspondence between Google and biology is not an analogy; it's a fact of life. Nucleic acid sequences are already being linked, via Google, to protein structures, and direct translation will soon be under way.

So much for the address limitation. The other limitation of which von Neumann was acutely aware was the language limitation, that a formal language based on precise logic can only go so far amidst real-world noise. "The message-system used in the nervous system . . . is of an essentially statistical character," he explained in 1956, just before he died. "In other words, what matters are not the precise positions of definite markers, digits, but the statistical characteristics of their occurrence . . . Whatever language the central nervous system is using, it is characterized by less logical and arithmetical depth than what we are normally used to [and] must structurally be essentially different from those languages to which our common experience refers." Although Google runs on a nutrient medium of von Neumann processors, with multiple layers of formal logic as a base, the higher-level meaning is essentially statistical in character. What connects where, and how frequently, is more important than the underlying code that the connections convey.

As Richard Foreman so beautifully describes it, we've been pounded into instantly available pancakes, becoming the unpredictable but statistically critical synapses in the whole Gödel-to-Google net. Does the resulting mind (as Richardson would have it) belong to us? Or does it belong to something else?

Turing proved that digital computers are able to answer most—but not all—problems that can be asked in unambiguous terms. They may, however, take a very long time to produce an answer (in which case you build faster computers) or it may take a very long time to ask the question (in which case you hire more programmers). This has worked surprisingly well for more than sixty years.

Most of real life, however, inhabits the third sector of the computational universe: where finding an answer is easier than defining the question. Answers are, in principle, computable, but, in practice, we are unable to ask the questions in unambiguous language that a computer can understand. It's easier to draw something that looks like a cat than to describe what, exactly, makes something look like a cat. A child scribbles indiscriminately, and eventually something appears that happens to resemble a cat. A solution finds the problem, not the other way around. The world starts making sense, and the meaningless scribbles are left behind.

"An argument in favor of building a machine with initial randomness is that, if it is large enough, it will contain every network that will ever be required," advised Turing's assistant, cryptanalyst Irving J. Good, in 1958. Random networks (of genes, of computers, of people) contain solutions, waiting to be discovered, to problems that need not be explicitly defined. Google has answers to questions no human being may ever be able to ask.

Operating systems make it easier for human beings to operate computers. They also make it easier for computers to operate human beings. (Resulting in Richard Foreman's "pancake effect.") These views are complementary, just as the replication of genes helps reproduce organisms, while the reproduction of organisms helps replicate genes. Same with search engines. Google allows people with questions to find answers. More importantly, it allows answers to find questions. From the point of view of the network, that's what counts. For obvious reasons, Google avoids the word "operating system." But if you are ever wondering what an operating system for the global computer might look like (or a true AI), a primitive but fully metazoan system like Google is the place to start.

Richard Foreman asked two questions. The answer to his first question is no. The answer to his second question is yes.

17.

The Age of the Informavore

Frank Schirrmacher

Copublisher, Frankfurter Allgemeine Zeitung *(FAZ); author,*
Das Methusalem-Komplott ("The Methusaleh Conspiracy")

The question I am asking myself arose through work and through discussion with other people, and especially watching other people, watching them act and behave and talk. It was how technology—the Internet and the modern systems—has now apparently changed human behavior, the way humans express themselves, and the way humans think in real life.

We are apparently now in a situation where modern technology is changing the way people behave, people talk, people react, people think, and people remember. And you encounter this not only in a theoretical way, but when you meet people, when suddenly people start forgetting things, when suddenly people depend on their gadgets and other stuff to remember certain things. This is the beginning; it's just an experience. But if you think about it and you think about your own behavior, you suddenly realize that something fundamental is going on. There is one comment on *Edge* that I love which is in Daniel Dennett's response to the 2007 annual question: He writes that we have a population explosion of ideas, but not enough brains to cover them.

As we know, information is fed by attention, so we have not enough attention, not enough food for all this information. And, as we know—this is the old Darwinian thought, the moment when Darwin started reading Malthus—when you have a conflict between a population explosion and not enough food, then Darwinian selection starts. And Darwinian systems start to change situations. And so what interests me is that we are, because we have the Internet, now entering a phase where Darwinian structures—where Darwinian dynamics, Darwinian selection—apparently attack ideas themselves: what to remember, what not to remember, which idea is stronger, which idea is weaker.

Here European thought is quite interesting, our whole history of thought, especially in the 18th, 19th, and 20th centuries, starting from Kant to Nietzsche. Hegel, for example, in the 19th century, where you said which thought, which thinking succeeds and which one doesn't. We have phases in the 19th century when you could have chosen either way. You could have gone the way of Schelling, for example, the German philosopher, which was totally different from that of Hegel. And so this question of what survives, which idea survives and which idea drowns, which idea starves to death, is something that, in our whole system of thought, is very, very well known and is quite an issue. And now we encounter this structure, this phenomenon, in everyday thinking.

It's the question: What is important, what is not important? What is important to know? Is this information important? Can we still decide what is important? And it starts with this absolutely normal, everyday news. But now you encounter, at least in Europe, a lot of people who think, what in my life is important, what isn't important, what is the information of my life? And some of them say, well, it's on Facebook. And others say, well, it's on my blog. And apparently for many people it's very hard to say it's somewhere in their life, in their lived life.

Of course, everybody knows we have a revolution, but we are now really entering the cognitive revolution of it all. In Europe, and in America, too—and it's not by chance—we have a crisis of all the systems that somehow are linked to either thinking or to knowledge. It's the publishing companies, it's the newspapers, it's the media, it's TV. But it's as well the university, and the whole school system, where it is not a normal crisis of too few teachers, too many pupils, or whatever; too small universities, too big universities.

Now, it's totally different. When you follow the discussions, there's the question of what to teach, what to learn, and how to learn. Even for universities and schools, suddenly they are confronted with the question, How can we teach? What is the brain actually taking? And then there are the problems that we have with attention deficit and all that, which are reflections and, of course, results, in a way, of the technical revolution.

Gerd Gigerenzer, whom I find a fascinating thinker, put it in such a way that thinking itself somehow leaves the brain and uses a platform outside of the human body. And that's the Internet, and it's the cloud. And very soon we will have the brain in the cloud. And this raises the question of the importance of thoughts. For centuries, what was important for me was decided in my brain. But now, apparently, it will be decided somewhere else.

The European point of view, with our history of thought and all our idealistic tendencies, is that now you can see—because they didn't know that the Internet would be coming, in the fifties or sixties or seventies—that the whole idea of the Internet somehow was built in the brains, years and decades before it actually was there, in all the different sciences. The computer—Gigerenzer wrote a great essay about that—at first was somehow isolated, it was in the military, in big laboratories, and so on. And then the moment came in the seventies and of course in the eighties when the computer was spread around, and every doctor, every household had a computer. Suddenly, the metaphors that were built in the fifties, sixties, and seventies had their triumph. And so people had to use the computer. As they say, the computer is the last metaphor for the human brain; we don't need any more. It succeeded because the tool shaped the thought when it was there, but all the thinking, like in brain sciences and all the others, had already happened, in the seventies, sixties, fifties even.

But the interesting question is, of course, the Internet—I don't know if they really expected the Internet to evolve the way it did. I've read books from the nineties in which they still didn't really know that it would be as huge as it is. And, of course, nobody predicted Google at that time.

Now, what I find interesting is that if you see the computer and the Web, and all this, under the heading of "the new technologies," we have, in the late 19th century, this big discussion about the human motor. The new machines in the late 19th century required that the muscles of the human being should be adapted to the new machines. Especially in Austria and Germany, we have this new thinking where people said, first of all, we have to change muscles. The term "calories" was invented in the late 19th century, in order to optimize the human workforce.

Now, in the 21st century, you have all the same issues, but with the brain. What was once the adaptation of muscles to the machines is now under the heading of multitasking—which is quite a problematic issue. The human muscle in the head, the brain, has to adapt. And, as we know from just very recent studies, it's very hard for the brain to adapt to multitasking, which is only one issue. And again with calories and all that, I think it's very interesting, the concept—again, Daniel Dennett and others said it—the concept of the informavores, the human being as somebody eating in-

Frank Schirrmacher

formation. So you can, in a way, see that the Internet and that the information overload we are faced with at this very moment has a lot to do with food chains, has a lot to do with the food you eat or do not eat—with food that has many calories and doesn't do you any good, and with food that is very healthy and is good for you.

The tool is not only a tool; it shapes the human who uses it. We always have the concept—first you have the theory, then you build the tool, and then you use the tool. But the tool itself is powerful enough to change the human being. God as the clockmaker, I think you said. Then in the Darwinian times, God was an engineer. And now He, of course, is the computer scientist and a programmer. What is interesting, of course, is that the moment neuroscientists and others used the computer, the tool of the computer, to analyze human thinking, something new started.

The idea that thinking itself can be conceived in technical terms is quite new. Even in the thirties, of course, you had all these metaphors for the human body, even for the brain; but, for thinking itself, this was very, very late. Even in the sixties, it was very hard to say that thinking is like a computer.

Edge published, years ago, a very interesting talk with Patty Maes on "Intelligence Augmentation," when she was one of the first who invented these intelligent agents. And there, you and Jaron Lanier, and others, asked the question about the concept of free will. And she explained it and it wasn't that big an issue, of course, because it was just intelligent agents like the ones we know from Amazon and others. But now, entering real-time Internet and all the other possibilities in the near future, the question of predictive search and others, of determinism, becomes much more interesting. The question of free will, which always was a kind of theoretical question—even very advanced people said,

Well, we declare there is no such thing as free will, but we admit that people, during their childhood, will have been culturally programmed so they believe in free will.

But now, you have a generation—in the next evolutionary stages, the child of today—that is adapted to systems such as the iTunes "Genius," which not only knows which book or which music file they like, but goes further and further in predicting certain things, like predicting whether the concert I am watching tonight is good or bad. Google will know it beforehand, because they know how people talk about it.

What will this mean for the question of free will? Because, in the bottom line, there are, of course, algorithms that analyze or that calculate certain predictabilities. And I'm wondering if the comfort of free will or not free will would be a very, very tough issue of the future. At this very moment, we have a new government in Germany; they are just discussing what kind of effect this will have on politics. And one of the issues, which of course at this very moment seems to be very isolated, is the question of how to predict certain terroristic activities, which they could do, from blogs—as you know, in America, you have the same thing. But this can go further and further.

The question of prediction will be the issue of the future and such questions will have impact on the concept of free will. We are now confronted with theories by psychologist John Bargh and others who claim there is no such thing as free will. This kind of claim is a very big issue here in Germany, and it will be a much more important issue in the future than we think today. The way we predict our own life, the way we are predicted by others, through the cloud, through the way we are linked to the Internet, will impact every aspect of our lives. And, of course, this will play out in the workforce—the new German government seems to be

Frank Schirrmacher

very keen on this issue, to at least prevent the worst impact on people, on workplaces.

It's very important to stress that we are not talking about cultural pessimism. What we are talking about is that a new technology that is a brain technology, or to put it this way, that is a technology having to do with intelligence—this new technology now clashes in a very real way with the history of thought in the European way of thinking.

Unlike America, in Germany we had a party for the first time in the last elections that totally comes out of the Internet. They are called the Pirates. In their beginnings they were computer scientists concerned with questions of copyright and all that. But it's now much, much more. In the recent election, out of the blue, they received 2 percent of the votes, which is a lot for a new party that only exists on the Internet. And the voters were 30, 40, 50 percent young males. Many, many young males. They're all very keen on new technologies. Of course, they are computer kids and all that. But this party, now, for the first time, reflects the way that we know, theoretically, in a very pragmatic and political way. For example, one of the main issues, as I just described, the question of the adaptation of muscles to modern systems, either in the brain or in the body, is a question of the digital Taylorism.

As far as we can see, I would say, we have three important concepts of the 19th century, which somehow come back in a very personalized way, just like you have a personalized newspaper. The first is Darwinism, the whole question. And, in a very real sense, look at the problem with Google and the newspapers. Darwinism, but as well the whole question of who survives in the Net, in the thinking, who gets more traffic, who gets less traffic, and so on. Second, you have the concept of communism, which comes back to the question of free, the question that people work for free. And not only those people who sit at home and write blogs, but also many people in publishing companies, at newspapers, do a lot of things for free or offer them for free. And then third, of course, Taylorism, which is a nonissue, but we now have the digital Taylorism, but with an interesting switch. At least in the 19th and the early 20th century, you could still make others responsible for your own deficits in that you could say, well, this is just really terrible, it's exhausting, and it's not human, and so on.

Now, look at the concept, for example, of multitasking, which is a real problem for the brain. You don't think that others are responsible for it, but you meet many people who say, well, I am not really good at it, and it's my problem, and I forget, and I am just overloaded by information. What I find interesting is that three huge political concepts of the 19th century come back in a totally personalized way, and that we now, for the first time, have a political party—a small political party, but it will in fact influence the other parties—who address this issue, again, in this personalized way.

It's a kind of catharsis, this Twittering, and so on. But now, of course, this kind of information conflicts with many other kinds of information. And, in a way, one could argue—I know that was the case with Iran—that maybe the future will be that the Twitter information about an uproar in Iran competes with the Twitter information of Ashton Kutcher, or Paris Hilton, and so on. The question is to understand which is important. What is important, what is not important is something very linear, it's something that needs time, at least the structure of time. Now, you have simultaneity, you have everything happening in real time. And this impacts politics in a way that might be considered good, but could also be bad.

Because suddenly it's gone again. And the next piece of information, and the next piece of information, and now—and this is something that, again, has very much to do with the concept of the European self, to take oneself seriously, and so on—now, as Google puts it, they say, if I understand it rightly, all these webcams and cell phones are full of information. There are photos, there are videos, whatever. And they all should be, if people want it, shared. And all the thoughts expressed in any university, at this very moment, there could be thoughts we really should know. I mean, in the 19th century, it was not possible. But maybe there is one student who is much better than any of the thinkers we know. So we will have an overload of all this information, and we will be dependent on systems that calculate, that make the selection of this information.

And, as far as I can see, political information somehow isn't distinct from it. It's the same issue. It's a question of whether I have information from my family on the iPhone, or whether I have information about our new government. And so this incredible amount of information somehow becomes equal, and very, very personalized. And you have personalized newspapers. This will be a huge problem for politicians. From what I hear, they are now very interested in, for example, Google's page rank; in the question how, with mathematical systems, you can, for example, create information cascades as a kind of artificial information overload. And, as you know, you can do this. And we are just not prepared for that. It's not too early. In the last elections we, for the first time, had blogs, where you could see they started to create information cascades, not only with human beings, but as well with Web robots and other stuff. And this is, as I say, only the beginning.

Germany still has a very strong antitechnology movement, which is quite interesting insofar as you can't really say it's left-wing or right-wing. As you know, very right-wing people, in German history especially, were very antitechnology. But it changed a lot. And why it took so long, I would say, has to do with demographic reasons. We are in an aging society, and the generation that is now forty or fifty, in Germany, had their children very late. The whole evolutionary change, through the new generation—first, they are fewer, and then, they came later. It's not like in the sixties and seventies, with Warhol. And the fifties. These were young societies. It happened very fast. We took over all these interesting influences from America very, very fast, because we were a young society. Now, somehow it really took a longer time, but now it is certain we are entering, for demographic reasons, the situation where a new generation—as you see with the Pirates as a party, they're a new generation—that grew up with modern systems, with modern technology, they are now taking the stage and changing society.

What did Shakespeare, and Kafka, and all these great writers—what actually did they do? They translated society into literature. And of course, at that stage, society was something very real, something that you could see. And they translated modernization into literature. Now we have to find people who translate what happens on the level of software. At least for newspapers, we should have sections that review software in a different way, at least the structures of software.

One must say, all the big software companies are American, except SAP. But Google and all these others, they are American companies. I would say we weren't very good at inventing. We are not very good at getting people to study computer science and other things. And I must say—and this is not meant as flattery of America or whomever—what I really miss is that we don't have this type of computationally minded intellectual—though it started in Germany once, decades ago—such as Danny Hillis and

Frank Schirrmacher

other people who participate in a kind of intellectual discussion, even if only a happy few read and react to it. Not many German thinkers have adopted this kind of computational perspective.

The ones who do exist have their own platform and actually created a new party. This is something we are missing, because there has always been a kind of an attitude of arrogance toward technology. For example, I am responsible for the entire cultural sections and science sections of *FAZ*. And we publish reviews about all these wonderful books on science and technology, and that's fascinating and that's good. But, in a way, the really important texts—which somehow write our life today and which are, in a way, the stories of our life—are, of course, the software. And these texts weren't reviewed. We should have found ways of transcribing what happens on the software level much earlier—like Patty Maes or others, just to write it, to rewrite it in a way that people understand what it actually means. I think this is a big lack.

We are just beginning to look at this in Germany. And we are looking for people—it's not very many people—who have the ability to translate that. It needs to be done because that's what makes us who we are. You will never really understand in detail how Google works because you don't have access to the code. They don't give you the information. But just think of George Dyson's essay, which I love, "Turing's Cathedral" [see chapter 6]. This is a very good beginning. He absolutely has a point. It is today's version of the kind of cathedral we would be entering if we lived in the 12th century. It's incredible that people are building this cathedral of the digital age. And as he points out, when he visited Google, he saw all the books they were scanning, and noted that they said they are not scanning these books for humans to read, but for the artificial intelligence to read.

Who are the big thinkers here? In Germany, at least for my work, there are a couple of key figures. One of them is Gerd Gigerenzer, who is somebody who is absolutely—I would say he is actually avant-garde, at this very moment, because what he does is he teaches heuristics. And from what we see, we have an amputation of heuristics, through the technologies, as well. People forget certain heuristics. It starts with a calculation, because you have the calculator, but it goes much further. And you will lose many more rules of thumb in the future because the systems are doing that, Google and all the others. So Gigerenzer, in his thinking—and he has the Max Planck Institute now—on risk assessment, as well, is very, very important. You could link him, in a way, actually to Nassim Taleb, because again here you have the whole question of risk assessment, the question of looking back, looking into the future.

Very important in literature, still, though he is seventy years old, eighty years old, is of course Hans Magnus Enzensberger. Peter Sloterdijk is a very important philosopher; a kind of literary figure, but he is important. But then you have, not unlike in the 19th or 20th century, many leading figures. We have quite interesting people, at this very moment, in law, which is very important for discussions of copyright and all that. But regarding conversations of new technologies and human thought, they, at this very moment, don't really take place in Germany.

There are European thinkers who have cult followings—Slavoj Žižek, for example. Ask any intellectuals in Germany, and they will tell you Žižek is just the greatest. He's a kind of communist; he considers himself Stalinistic, even. But this is, of course, all labels. Wild thinkers. Europeans, at this very moment, love wild thinkers.

Frank Schirrmacher

BOOKS BY JOHN BROCKMAN

CULTURE
ISBN 978-0-06-202313-1 (paperback)

THE MIND
ISBN 978-0-06-202584-5 (paperback)

IS THE INTERNET CHANGING THE WAY YOU THINK?
ISBN 978-0-06-202044-4 (paperback)

THIS WILL CHANGE EVERYTHING
ISBN 978-0-06-189967-6 (paperback)

WHAT HAVE YOU CHANGED YOUR MIND ABOUT?
ISBN 978-0-06-168654-2 (paperback)

WHAT ARE YOU OPTIMISTIC ABOUT?
ISBN 978-0-06-143693-2 (paperback)

WHAT IS YOUR DANGEROUS IDEA?
ISBN 978-0-06-121495-0 (paperback)

WHAT WE BELIEVE BUT CANNOT PROVE
ISBN 978-0-06-084181-2 (paperback)